Biometric Recognition
CHALLENGES AND OPPORTUNITIES

Joseph N. Pato and Lynette I. Millett, *Editors*

Whither Biometrics Committee

Computer Science and Telecommunications Board

Division on Engineering and Physical Sciences

NATIONAL RESEARCH COUNCIL
OF THE NATIONAL ACADEMIES

THE NATIONAL ACADEMIES PRESS
Washington, D.C.
www.nap.edu

THE NATIONAL ACADEMIES PRESS   500 Fifth Street, N.W.   Washington, DC 20001

NOTICE: The project that is the subject of this report was approved by the Governing Board of the National Research Council, whose members are drawn from the councils of the National Academy of Sciences, the National Academy of Engineering, and the Institute of Medicine. The members of the committee responsible for the report were chosen for their special competences and with regard for appropriate balance.

Support for this project was provided by the Defense Advanced Research Projects Agency (Award No. N00174-03-C-0074) and by the Central Intelligence Agency and the Department of Homeland Security with assistance from the National Science Foundation (Award No. IIS-0344584). Any opinions expressed in this material are those of the authors and do not necessarily reflect the views of the agencies and organizations that provided support for the project.

International Standard Book Number-13: 978-0-309-14207-6
International Standard Book Number-10: 0-309-14207-5

Copies of this report are available from

The National Academies Press
500 Fifth Street, N.W., Lockbox 285
Washington, DC 20055
800/624-6242
202/334-3313 (in the Washington metropolitan area)
http://www.nap.edu

Printed in the United States of America

THE NATIONAL ACADEMIES
Advisers to the Nation on Science, Engineering, and Medicine

The **National Academy of Sciences** is a private, nonprofit, self-perpetuating society of distinguished scholars engaged in scientific and engineering research, dedicated to the furtherance of science and technology and to their use for the general welfare. Upon the authority of the charter granted to it by the Congress in 1863, the Academy has a mandate that requires it to advise the federal government on scientific and technical matters. Dr. Ralph J. Cicerone is president of the National Academy of Sciences.

The **National Academy of Engineering** was established in 1964, under the charter of the National Academy of Sciences, as a parallel organization of outstanding engineers. It is autonomous in its administration and in the selection of its members, sharing with the National Academy of Sciences the responsibility for advising the federal government. The National Academy of Engineering also sponsors engineering programs aimed at meeting national needs, encourages education and research, and recognizes the superior achievements of engineers. Dr. Charles M. Vest is president of the National Academy of Engineering.

The **Institute of Medicine** was established in 1970 by the National Academy of Sciences to secure the services of eminent members of appropriate professions in the examination of policy matters pertaining to the health of the public. The Institute acts under the responsibility given to the National Academy of Sciences by its congressional charter to be an adviser to the federal government and, upon its own initiative, to identify issues of medical care, research, and education. Dr. Harvey V. Fineberg is president of the Institute of Medicine.

The **National Research Council** was organized by the National Academy of Sciences in 1916 to associate the broad community of science and technology with the Academy's purposes of furthering knowledge and advising the federal government. Functioning in accordance with general policies determined by the Academy, the Council has become the principal operating agency of both the National Academy of Sciences and the National Academy of Engineering in providing services to the government, the public, and the scientific and engineering communities. The Council is administered jointly by both Academies and the Institute of Medicine. Dr. Ralph J. Cicerone and Dr. Charles M. Vest are chair and vice chair, respectively, of the National Research Council.

www.national-academies.org

WHITHER BIOMETRICS COMMITTEE

JOSEPH N. PATO, Hewlett-Packard Company, *Chair*
BOB BLAKLEY, Gartner
JEANETTE BLOMBERG, IBM Almaden Research Center
JOSEPH P. CAMPBELL, Massachusetts Institute of Technology, Lincoln
  Laboratory
GEORGE T. DUNCAN, Carnegie Mellon University
GEORGE R. FISHER, Prudential-Wachovia (retired)
STEVEN P. GOLDBERG,[1] Georgetown University Law Center
PETER T. HIGGINS, Higgins & Associates, International
PETER B. IMREY, Cleveland Clinic and Case Western Reserve
  University
ANIL K. JAIN, Michigan State University
GORDON LEVIN, The Walt Disney World Company
LAWRENCE D. NADEL, Noblis
JAMES L. WAYMAN, San Jose State University

Staff

LYNETTE I. MILLETT, Senior Program Officer

---

[1] Steven P. Goldberg died on August 26, 2010.

Preface

In a variety of government and private domains biometric recognition is being promoted as a technology that can help identify terrorists, provide better control of access to physical facilities and financial accounts, and increase the efficiency of access to services and their utilization. Biometric recognition has been applied to identification of criminals, patient tracking in medical informatics, and the personalization of social services, among other things. In spite of substantial effort, however, there remain unresolved questions about the effectiveness and management of systems for biometric recognition, as well as the appropriateness and societal impact of their use. Moreover, the general public has been exposed to biometrics largely as high-technology gadgets in spy thrillers or as fear-instilling instruments of state or corporate surveillance in speculative fiction.

Now, at the beginning of the second decade of the twenty-first century, biometric technologies appear poised for broader use. Increased concerns about national security and the tracking of individuals as they cross borders have caused passports, visas, and border-crossing records to be linked to biometric data. A focus on fighting insurgencies and terrorism has led to the military deployment of biometric tools to enable recognition of individuals as friend or foe. Commercially, finger-imaging sensors, whose cost and physical size have been reduced, now appear on many laptop personal computers, handheld devices, mobile phones, and other consumer devices.

In 2001 the Computer Science and Telecommunications Board (CSTB) of the National Research Council (NRC) formed a committee whose 2003 report *Who Goes There? Authentication Through the Lens of Privacy*, considered several authentication technologies, one of which was biometrics. After the publication of that report, the CSTB held several discussions with various federal agencies interested in biometrics. Jonathon Phillips (then at the Defense Advanced Research Projects Agency (DARPA)), Gary Strong (then at the Department of Homeland Security (DHS)), and Andrew Kirby (of the Central Intelligence Agency (CIA)) actively participated in the discussions and helped to move them forward. The discussions resulted in agreement to undertake this comprehensive assessment of biometrics (see Appendix C for the project's original statement of task). Funding for the project was obtained from DARPA and from the CIA and the DHS with assistance from the National Science Foundation. The Whither Biometrics Committee was formed to conduct the study.

The Whither Biometrics Committee consisted of 13 members[1] from industry and academia who are experts in different aspects of distributed systems, computer security, biometrics (of various flavors), systems engineering, human factors, the law, and statistics, as well as in computer science and engineering (see Appendix A for committee and staff biographies).

Early in the study the committee organized a public workshop. Held on March 15 and 16, 2005, in Washington, D.C., the workshop was attended by members of industry, government, and academia and reported on by the committee in *Summary of a Workshop on the Technology, Policy, and Cultural Dimensions of Biometric Systems*.[2] In the course of the study, inputs were gathered on the challenges, capabilities, and requirements of biometric systems as well as related policy and social questions. This report draws on what was learned at the workshop and in subsequent briefings to the committee.

The report makes two main points. First, developers and analysts of biometric recognition systems must bear in mind that such systems are complex and need to be addressed as such. Second, biometric recognition is an inherently probabilistic endeavor. The automated recognition of individuals offered by biometric systems must be tempered by an awareness of the uncertainty associated with that recognition. Uncertainty arises in numerous ways in biometric systems, including from poor or incomplete

---

[1] Delores Etter was originally a member of the committee but resigned when she was appointed Assistant Secretary of Research, Development, and Acquisition for the U.S. Navy.

[2] National Research Council, *Summary of a Workshop on the Technology, Policy, and Cultural Dimensions of Biometric Systems*, Kristen Batch, Lynette I. Millett, and Joseph N. Pato, eds., The National Academies Press, Washington, D.C. (2006).

understanding of the distinctiveness and stability of the traits measured by biometric systems; the difficulty of characterizing the probability that an imposter will attack the system; and even the attitudes of the subjects using the systems—subjects who may have become conditioned by fictional depictions to expect, or even fear, that recognition will be perfect. Consequently, even when the technology and the system it is embedded in are behaving as designed, there is inevitable uncertainty and risk of error. The probabilistic nature of biometric systems also means that the measured characteristics of the population of intended users (those the system is designed to recognize) matter and affect design and implementation choices.

This report elaborates on these themes in detail and is aimed at a broad audience, including policy makers, developers, and researchers. For policy makers, it seeks to provide a comprehensive assessment of biometric recognition that examines current capabilities, future possibilities, and the role of government in technology and system development. For developers and researchers, the report's goals are to articulate challenges posed by understanding and developing biometric recognition systems and to point out opportunities for research. Building on CSTB's work on authentication technologies and privacy, it explores the technical and policy challenges associated with the development, evaluation, and use of biometric technologies and systems that incorporate them.

The committee members brought different and complementary perspectives to their efforts as they deliberated and solicited input from a number of other experts. The committee held six plenary meetings, including the workshop. It thanks the many individuals who contributed, including the project sponsors that enabled this activity. The committee also conducted three site visits, one to the Boston Police Department's Identification Center, one to the U.S. Naval Academy, and another to Walt Disney World. The committee thanks those who came and briefed the committee at those meetings and site visits: Andrew Kirby, Joseph Kielman, John Atkins, Martin Herman, Duane Blackburn, Jean-Christophe Fondeur, James Matey, Sharath Pankanti, Jonathon Phillips, David Scott, George Doddington, Michele Freadman, Patrick Grother, Austin Hicklin, Nell Sedransk, Tora Bikson, David Kaye, Lisa Nelson, Peter Swire, Joseph Atick, Rick Lazarick, Tony Mansfield, Marek Rejman-Greene, Valorie Valencia, Cynthia Musselman, William Casey, Patty Cogswell, Neal Latta, K.A. Taipale, John Woodward, Jim Dempsey, Ari Schwartz, Michael Cherry, Mike Labonge, Richard Nawrot, Diane Ley, John Schmitt, Michael Wong, Vance Bjorn, Betty LaCrois, Ken Fong, Joseph Dahlbeck, Dennis Treece, and Lynne Hare. It appreciates briefers' willingness to answer the questions they were asked and is grateful for their insights. Additional information was garnered from reviewing the published literature and obtaining informal input at various conferences and other meetings. Input was also derived from committee members during the course of their professional activities outside the committee's work.

It is with great sadness that we mourn the passing of our colleague and fellow committee member Steven Goldberg, who died just prior to this report's publication. He was a valued member of our study team. His insights on science and the law and his collegial and constructive approach to interdisciplinary work are greatly missed.

We thank the sponsors who enabled this project, the reviewers whose constructive criticism improved the report, and the editor Liz Fikre for her help in refining the final draft of the report. The committee is grateful to the CSTB staff members whose work has made this report possible. The committee thanks Jon Eisenberg for his extensive helpful feedback throughout the process, Margaret Huynh for impeccable coordination of logistics, Kristen Batch for her work in assisting with our earlier workshop report, and Ted Schmitt, who helped structure early drafts of the final report. Finally, we thank Lynette Millett, Senior Program Officer, who has ably guided this project as study director from its inception and was essential to completing our work.

Joseph N. Pato, *Chair*
Whither Biometrics Committee

Acknowledgment of Reviewers

This report has been reviewed in draft form by individuals chosen for their diverse perspectives and technical expertise, in accordance with procedures approved by the National Research Council's Report Review Committee. The purpose of this independent review is to provide candid and critical comments that will assist the institution in making its published report as sound as possible and to ensure that the report meets institutional standards for objectivity, evidence, and responsiveness to the study charge. The review comments and draft manuscript remain confidential to protect the integrity of the deliberative process. We wish to thank the following individuals for their review of this report:

Michael F. Angelo, Net IQ,
Ming Hsieh, Cogent Systems, Inc.,
Stephen Kent, BBN Technologies,
Sara Kiesler, Carnegie Mellon University,
Herbert Levinson, Transportation Consultant,
Steven Lipner, Microsoft Corporation,
Helen Nissenbaum, New York University,
Louise Ryan, Harvard School of Public Health,
Michael Saks, Arizona State University, and
Valorie Valencia, Authenti-Corp.

Although the reviewers listed above have provided many constructive comments and suggestions, they were not asked to endorse the conclusions or recommendations, nor did they see the final draft of the report before its release. The review of this report was overseen by Robert F. Sproull of Oracle Corporation. Appointed by the National Research Council, he was responsible for making certain that an independent examination of this report was carried out in accordance with institutional procedures and that all review comments were carefully considered. Responsibility for the final content of this report rests entirely with the authoring committee and the institution.

Contents

APPENDIXES

Summary

Biometrics is the automated recognition of individuals based on their behavioral and biological characteristics. It is a tool for establishing confidence that one is dealing with individuals who are already known (or not known)—and consequently that they belong to a group with certain rights (or to a group to be denied certain privileges). It relies on the presumption that individuals are physically and behaviorally distinctive in a number of ways. Figure S.1 illustrates the basic operations of a recognition process.

Biometric systems are used increasingly to recognize individuals and regulate access to physical spaces, information, services, and to other rights or benefits, including the ability to cross international borders. The motivations for using biometrics are diverse and often overlap. They include improving the convenience and efficiency of routine access transactions, reducing fraud, and enhancing public safety and national security. Questions persist, however, about the effectiveness of biometric systems as security or surveillance mechanisms, their usability and manageability, appropriateness in widely varying contexts, social impacts, effects on privacy, and legal and policy implications.

The following are the principal conclusions of this study:

- Human recognition systems are inherently probabilistic, and hence inherently fallible. The chance of error can be made small but not eliminated. System designers and operators should anticipate and plan for the occurrence of errors, even if errors are expected to be infrequent.
- The scientific basis of biometrics—from understanding the distributions of biometric traits within given populations to how humans

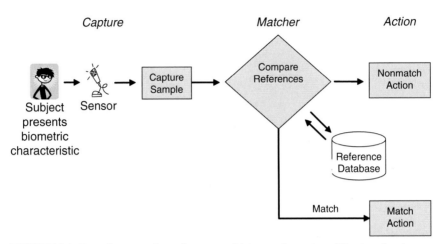

FIGURE S.1 Sample operation of a general biometric system. The two basic operations performed by a general biometric system are the capture and storage of enrollment (reference) biometric samples and the capture of new biometric samples and their comparison with corresponding reference samples (matching). This figure depicts the operation of a generic biometric system although some systems will differ in their particulars. The primary components for the purposes of this discussion are "capture," where the sensor collects biometric data from the subject to be recognized; the "reference database," where previously enrolled subjects' biometric data are held; the "matcher," which compares presented data to reference data in order to make a recognition decision; and "action," where the system recognition decision is revealed and actions are undertaken based on that decision.

interact with biometric systems—needs strengthening particularly as biometric technologies and systems are deployed in systems of national importance.

• Biometric systems incorporate complex definitional, technological, and operational choices, which are themselves embedded in larger technological and social contexts. Thus, systems-level considerations are critical to the success of biometric systems. Analyses of biometric systems' performance, effectiveness, trustworthiness, and suitability should take a broad systems perspective.

• Biometric systems should be designed and evaluated relative to their specific intended purposes and contexts rather than generically. Their effectiveness depends as much on the social context as it does on the underlying technology, operational environment, systems engineering, and testing regimes.

• The field of biometrics would benefit from more rigorous and comprehensive approaches to systems development, evaluation, and interpretation. Presumptions and burdens of proof arising from biometric recognition should be based on solid, peer-reviewed studies of the performance of biometric recognition mechanisms.

FUNDAMENTALS OF BIOMETRIC RECOGNITION AND HUMAN INDIVIDUAL DISTINCTIVENESS

Biometric recognition systems are inherently probabilistic, and their performance needs to be assessed within the context of this fundamental and critical characteristic. Biometric recognition involves matching, within a tolerance of approximation, of observed biometric traits against previously collected data for a subject. Approximate matching is required due to the variations in biological attributes and behaviors both within and between persons.[1] Consequently, in contrast to the largely binary results associated with most information technology systems, biometric systems provide probabilistic results.

There are numerous sources of uncertainty and variation in biometric systems, including the following:

- *Variation within persons.* Biometric characteristics and the information captured by biometric systems may be affected by changes in age, environment, disease, stress, occupational factors, training and prompting, intentional alterations, sociocultural aspects of the situation in which the presentation occurs, changes in human interface with the system, and so on. As a result, each interaction of the individual with the system (at enrollment, identification, and so on) will be associated with different biometric information. Individuals attempting to thwart recognition for one reason or another also contribute to the inherent uncertainty in biometric systems.
- *Sensors.* Sensor age and calibration, how well the interface at any given time mitigates extraneous factors, and the sensitivity of sensor performance to variation in the ambient environment (such as light levels) all can play a role.
- *Feature extraction and matching algorithms.* Biometric characteristics cannot be directly compared but require stable and distinctive "features" to first be extracted from sensor outputs. Differences in feature extraction algorithms affect performance, with effects sometimes aggravated by requirements for achieving interoperability among proprietary systems. Differences between matching algorithms and comparison scoring mecha-

---

[1]For example, each finger of each person will generate a different fingerprint image every time it is observed due to presentation angle, pressure, dirt, moisture, different sensors, and so on. Thus each person can produce a large number of different impressions from a single finger—many of which will be close enough that good algorithms can match them to the correct finger source.

nisms, and how these interact with the preceding sources of variability of information acquired and features extracted, also contribute to variation in performance of different systems.

   • *Data integrity.* Information may be degraded through legitimate data manipulation or transformation or degraded and/or corrupted owing to security breaches, mismanagement, inappropriate compression, or some other means. It may also be inappropriately applied to a context other than the one for which it was originally created, owing to mission creep (for example, using the data collected in a domain purely for the sake of convenience in a domain that demands high data integrity) or inappropriate re-use of information (for instance, captured biometric information might be incorrectly assumed to be of greater fidelity when transferred to a system where higher fidelity is the norm).

   Many gaps exist in our understanding of the nature and extent of distinctiveness and stability of biometric traits across individuals and groups. No biometric characteristic is known to be entirely stable and distinctive across all groups. Biometric traits have fundamental statistical properties, distinctiveness, and differing degrees of stability under natural physiological conditions and environmental challenges, many aspects of which are not well understood, especially at large scales. Complicating matters, the underlying biological properties and distribution of biometric traits in a population are generally observed only through filters interposed by measurement processes and instruments and subsequent biometric feature extraction.

   Thus, the development of a science of human individual distinctiveness is essential to effective and appropriate use of biometric recognition. Better understanding of biometric traits in human beings could be gained by carefully designed data collection and analysis. The biological underpinnings of physical distinctiveness and the stability of many biometric characteristics under natural physiological conditions and environmental challenges require further justification from basic biological and empirical studies. Importantly, the underlying distinctiveness of a biometric trait cannot be assessed apart from an understanding of the stability, accuracy, and inherent variability of a given measure.

   Another fundamental characteristic of biometric recognition is that it requires decision making under uncertainty by both the automated recognition system and the human interpreters of its results. A biometric match represents not certain recognition but a probability of correct recognition, while a nonmatch represents a probability rather than a definitive conclusion that an individual is not known to the system. That is, some fraction of results from even the best-designed biometric system will be incorrect or indeterminate: both false matches and false nonmatches will occur. Moreover, assessing the validity of the match results, even given this inherent uncertainty, requires knowledge of the population of users who are presenting to the system—specifically, what proportions of those users should and should not match. Even very small probabilities of misrecognitions—the failure to recognize an enrolled individual or the recognition of one individual as another—can become operationally significant when an application is scaled to handle millions of recognition attempts. Thus, well-articulated processes for verification, mitigation of undesired outcomes, and remediation (for misrecognitions) are needed, and presumptions and burdens of proof should be designed conservatively, with due attention to the system's inevitable uncertainties.

Principle: Users and developers of biometric systems should recognize and take into account the limitations and constraints of biometric systems—especially the probabilistic nature of the underlying science, the current limits of knowledge regarding human individual distinctiveness, and the numerous sources of uncertainty in biometric systems.

BIOMETRIC SYSTEMS AND TRUSTWORTHINESS

Systems that perform biometric recognition exist within a constellation of other authentication and identification technologies and offer some distinct capabilities and challenges. Authentication technologies are typically based on one of three things: something the individual knows, such as a password; something the individual has, such as a physical key or secure token; and something the individual is or does.[2] Biometric technologies employ the last of these. Unlike password- or token-based systems, biometric systems can function without active input, user cooperation, or knowledge that the recognition is taking place.

Biometric systems, therefore, are not a general replacement for other authentication technologies, although combining biometric approaches with other methods can augment security in those applications where user cooperation can be inferred.

One important difference between biometric and other authentication technologies, such as tokens or passwords, is that these other technologies place trust in cooperative users, allowing them to produce what they possess or demonstrate what they know (through dependence on the user's safekeeping of a card or password). But these other forms of authentication do not protect against the sharing or transfer of the token or secret,

---

[2]Federal Information Processing Standards 48, "Guidelines on Evaluation of Techniques for Automated Personal Identification," was published in 1977 and was one of the first such treatments of authentication.

whereas biometric traits are tied to an individual[3]—specifically something an individual is or does.[4] Unintended disclosure of biometric data, however, may lead to more serious consequences or to consequences that are more difficult to remediate than the loss of a token or exposure of a password. Another important difference is that because they are probabilistic, biometric systems are particularly vulnerable to deliberate attempts to undermine confidence in their reliability, and discussions of probabilistic uncertainty can easily be twisted into a suggestion that biometric systems are unreliable.

Security challenges for biometric systems can be seen as stemming from two different views of such systems: (1) the use of biometric systems as a security mechanism to protect information systems or other resources and (2) vulnerabilities of the biometric system itself. First, it is necessary to determine if a biometric system is an appropriate component for the application at hand at all. One needs to specify the problem to be solved by a particular biometric system in order to adequately assess its effectiveness and deal with the consequences of deployment.[5] Conducting a threat analysis and developing threat models for the system that incorporates analysis of feasibility of threats against the resource being protected and against the system doing the protecting is an important component of understanding the problem. Decisions about whether and how to incorporate biometric approaches should consider their appropriateness and proportionality given the problem to be solved and the merits and risks of biometrics relative to other solutions[6] and need to be considered by the broader information security community as well as within the biometrics community.

Second, biometric systems (and not merely the resources they are protecting) are themselves vulnerable to attacks aimed at undermining their integrity and reliability. For password- or token-based systems, a breach can usually be remediated by issuing a new password or token.

---

[3]While it is possible to copy or mimic some biometric traits, it is generally more difficult to produce such a trait and present it to a supervised sensor than to share a password or token. If the system is unsupervised, an attacker may not need to spoof the trait physically; he might have a copy of the bit string or the reference, which would make such an attack no more difficult than compromising other forms of recognition.

[4]More precisely, biometric authentication is a binary hypothesis test where the hypothesis is that the biometric sample input matches—to a degree of certainty—the claimed biometric reference enrollment. The overall system then uses the matching results to accept or reject this hypothesis.

[5]See National Research Council, *Who Goes There? Authentication Through the Lens of Privacy* (2003) and *IDs—Not That Easy* (2002) for discussions of the need to understand the problem that a system is trying to solve in order to evaluate the system's effectiveness.

[6]For example, the problem of managing members' access to a local health club merits different kinds of analysis than does handling customs and immigration at a major international airport.

However, it is generally not possible to replace a biometric trait that has been compromised. This is complicated by the fact that the same biometric trait can be used by different systems, and weaknesses in one system could lead to the compromise of the biometric trait for use in another system. Furthermore, such traits are not secret—we expose them in the course of everyday life. For example, we leave fingerprints on many surfaces we touch, faces can be photographed, and voices can be recorded. However, it is as difficult for an impostor to grow a set of fingerprints matching those stolen as it is for the person they were stolen from to grow a new and different set. It is, accordingly, essential to validate that a trait presented to gain recognition truly belongs to the subject and is not being synthesized by an imposter. This often requires a human operator to observe the subject's presentation of the trait—which significantly constrains remote or distributed applications of biometrics. Automated verification that a living person is presenting what could conceivably be a synthesized artifact might be sufficient in some applications but would not substitute for human supervision where high degrees of confidence are required.

It is important to manage the trustworthiness of the entire process rather than focusing on evaluation of the proffered biometric characteristic. Systems using biometric recognition are typically designed with alternative procedures for use when a sensor fails or an individual lacks the biometric trait. Adversaries may attempt to force the system into failure modes to evade or accomplish recognition, implying that secondary screening procedures should be just as robustly designed as the main procedure. One potential way to improve recognition would be to use multiple biometric modalities and other demographic data to narrow the search space. This approach might have other advantages, such as expanding population coverage beyond that afforded by a single biometric and reducing vulnerability to spoofing attacks. It might have disadvantages, as well, including increasing the complexity and cost of the system. There are also issues related to the architecture and operation of multibiometrics systems as well as questions of how best to model such systems and then use the model to drive operational aspects. Understanding any statistical dependencies is critical when using multibiometrics.

TESTING, DESIGN, AND DEPLOYMENT

Although traditional biometrics testing tends to focus on the match performance for a test data set, experience from many domains suggests that process and quality control should be analyzed for the complete system life cycle. Methods used successfully for the study and improvement of systems in other fields such as manufacturing and medicine (for example, controlled observation and experimentation on operational systems guided by scientific principles and statistical design and monitoring) should be used in developing, maintaining, assessing, and improving biometric systems. One especially important lesson is that testing methods and results should be sufficiently open to allow independent assessment.

Although laboratory evaluations of biometric systems are highly useful for development and comparison, their results often do not reliably predict field performance. Operational testing and blind challenges of operational systems tend to give more accurate and usable results than developmental performance evaluations and operational testing in circumscribed and controlled environments. Although the international standards community has made progress in developing a coherent set of best practices for technology and scenario testing, guidelines for operational testing are still under development.[7] Designing a system and tests that can cope with ongoing data collection, particularly at scale, is a challenge making it difficult for a potential user of biometrics to determine how well a vendor's technology might operate in that user's applications or to measure improvements in the system's performance.

Principle: Efforts to determine best practices for testing and evaluating existing and new biometric systems should be sustained and expanded. Careful consideration should be given to making the testing process open, allowing assessment of results and quality measures by outside parties when appropriate. The evaluation of a system's effectiveness needs to take into account the purpose for which the system was developed and how well field conditions were matched.

It is essential to take a broad systems view when assessing the performance of biometric systems. Both enthusiasm for biometric recognition and concerns about it tend to focus narrowly on behavioral and biological characteristics, human interactions with biometric sensors, or how information collected will be used. Yet the effective use of biometrics involves more than simply engineering a system to provide these basic capabilities. Achieving automated recognition involves the proper functioning of a broader system with many elements, including the human sources of data, human operators of the system, the collection environment(s), biometric sensors, the quality of the system's various technological components, the human-sensor-environment interaction, biometric reference information databases and the quality and integrity of the data therein, the system's security and availability, the system's communications network(s), and the system's failure-handling and error-recovery processes.

---

[7]As of this writing, ISO/IEC Standard 19795-6 for operational testing is under development by ISO/IEC JTC1 SC37.

Successful deployments have good project management and definition of goals, alignment of biometric capabilities with the underlying need and operational environment, and a thorough threat and risk analysis. Failure is often rooted in a lack of clarity about the problem being addressed, lack of a viable business case, inappropriate application of biometrics where other technologies would work better, inappropriate choice of biometric technologies, insensitivity to user perceptions and usability requirements, inadequate support processes and infrastructure, and/or poor understanding of population issues among those to be recognized. User behavior, attitudes, and system usability contribute to misrecognitions, and how incorrect or indeterminate results are handled contributes to whether a system's goals are met.

The probabilistic nature of biometric systems makes them especially sensitive to how well exception mechanisms are implemented. In particular, the inevitable false matches, false nonmatches, and failures to enroll are likely to stress other portions of the system that have been put in place to compensate when such errors occur. Field error rates are likely to be higher than laboratory testing suggests, poor exception processes can negate benefits, and extrapolation of functions in one context to another context may be inappropriate.

Biometric systems should be designed to anticipate the development and adoption of new advances and standards, modularizing components that are likely to become obsolete, such as biometric sensors and matcher systems, so that they can be easily replaced. A life-cycle approach such as this requires understanding and taking into account the capabilities and limitations of biometric technologies and devices. Some of the factors that may compromise later use if systems are not backwards-compatible include degradation of data through transformations due to system interconnection or changes in technology and reuse of data in unanticipated applications. Exception policies, data quality threshold settings, and the consequences of false matches and false nonmatches may need adjustment over the life of a deployment, and provisions for such adjustments should be included in the system design. Training and outreach materials for a nonscientific audience are needed, along with strategies for dissemination to system operators. A life-cycle-oriented approach should also be flexible enough to manage the unexpected reactions of users, operators, or other stakeholders.

Principle: Best practices are needed for the design and development of biometric systems and the processes for their operation. To scale efficiently to mass applications, these best practices should include requirements for system usability, initial and sustained technical accuracy and system performance, appropriate exception handling, and consistency of adjudication at the system level. Best practices should allow for incorporation of scientific advances and be auditable throughout the life of the system.

System requirements can range widely depending on the user context, the application context, and the technology context. Issues related to the user context include motivations for using the system, users' awareness of their interactions with a system, and training and habituation to its use. Issues related to the application context include whether the system is supervised by human staff, whether it is being used to verify a positive recognition claim or a negative one, whether the population to be recognized is an open or closed group, and whether testing the claim requires one comparison or many. Issues related to the technology context include whether the environment (say, the lighting) is controlled, whether the system is covert or overt, passive or active (requiring interaction with the subject), how quickly users need to be processed, and the error rates required (based, for instance, on the consequence of errors). The issues related to these contexts should affect the system design, development, and deployment. In particular, the wide variety of options for a biometric system encompassed above make clear that the incorporation of biometrics in a system in and of itself says very little about the requirements or usage expectations of that system.

Principle: Requirements have critical implications for the design and development of human recognition systems and whether and how biometric technologies are appropriately employed. Requirements for systems can vary widely, and assessment and evaluation of the effectiveness of a given system need to take into account the problem and context it was intended to address.

SOCIAL, CULTURAL, AND LEGAL CONSIDERATIONS

Although biometric systems can be beneficial, the potentially lifelong association of biometric traits with an individual, their potential use for remote detection, and their connection with identity records may raise social, cultural, and legal concerns. When used in contexts where individuals are claiming enrollment or entitlement to a benefit, biometric systems could disenfranchise people who are unable to participate for physical, social, or cultural reasons. For these reasons, the use of biometrics—especially in applications driven by public policy, where the affected population may have little alternative to participation—merits careful oversight and public discussion to anticipate and minimize detrimental societal and individual effects and to avoid violating privacy and due process rights.

Social, cultural, and legal issues can affect a system's acceptance by users, its performance, or the decisions on whether to use it in the first place—so it is best to consider these explicitly in system design. Clearly, the behavior of those being enrolled and recognized can influence the accuracy and effectiveness of virtually any biometric system, and user behavior can be affected by the social, cultural, or legal context. Likewise, the acceptability of a biometric system depends on the social and cultural values of the participant populations. A careful analysis and articulation of these issues and their trade-offs can improve both acceptability and effectiveness. Moreover, the benefits arising from using a biometric system may flow to particular individuals or groups, sometimes only at the expense of others—for example, a building's owner might be more secure but at the cost of time and inconvenience to those who wish to enter the building—making calculating these trade-offs more difficult.

Fundamental to most social issues surrounding biometric recognition is the tight link between an individual's biometric traits and data record, which can have positive and negative consequences. These consequences can affect the disposition of a target population toward a particular application. The potential for disenfranchisement means that some could be excluded from the benefits of positive claim systems, including access to buildings and information or qualification for jobs or insurance. Policies and interfaces to handle error conditions such as failure to enroll or be recognized should be designed to gracefully avoid violating the dignity, privacy, or due process rights of the participants. In addition, the potential for abuse of power is a cause for concern. Many fear misuse of identification technology by authorities (from data compromise, mission creep, or use of a biometric for other than specified purposes). To be effective, biometric deployments need to take these fears seriously.

Some biometric systems are designed to recognize and track individuals without their knowledge. Covert identification has not been widely deployed, but its potential use raises deep concerns. Although the biometrics industry has at times dismissed such concerns, biometric systems could win broader acceptance if more attention were paid to the target community's cultural values.

Biometric recognition raises important legal issues of remediation, authority, and reliability, and, of course, privacy. The standard assumptions of the technologists who design new techniques, capabilities, and systems are very different from those embedded in the legal system. Legal precedent on the use of biometric technology is growing, with some key cases going back decades,[8] and other more recent cases[9] having raised serious questions about the admissibility of biometric evidence in court.

---

[8]Cases include *U.S. v. Dionisio* (U.S. Supreme Court, 1973) and *Perkey v. Department of Motor Vehicles* (California Supreme Court, 1986).

[9]Such as *Maryland v. Rose* (Maryland Circuit Court, 2007).

Remediation is one way of dealing with fraudulent use of biometrics (such as identity fraud or altering biometric reference data). Remediation also deals with individuals denied their due rights or access because of an incorrect match or nonmatch. Policy and law should not only address the perpetrators of fraud but also induce system owners to minimize misuse of biometric samples and to maximize appropriate monitoring of biometric sample presentation at enrollment and participation.

The reliability of biometric recognition is clouded by the presumption of near-infallibility promoted by popular culture. Such presumptions could make contesting improper identifications excessively difficult. Conversely, if all evidence must be up to the standards implied by certain popular culture phenomena, unreasonable difficulties could be faced in cases lacking sufficient resources or evidence to meet those standards.

The courts have sometimes taken the view that an individual's expectation of privacy is related to the ubiquity of a technical means, which implies that the legal status of challenges to biometric technologies could be affected by the commonality of their use.

Principle: Social, legal, and cultural factors can affect the acceptance and effectiveness of biometric systems and should be taken into account in system design, development, and deployment. Notions of proof related to biometric recognition should be based on solid, peer-reviewed studies of system accuracy under many conditions and for many persons reflecting real-world sources of error and uncertainty in those mechanisms. Pending scientific consensus on the reliability of biometric recognition mechanisms, a reasonable level of uncertainty should be acknowledged for biometric recognition. There may be a need for legislation to protect against the theft or fraudulent use of biometric systems and data.

ELEMENTS OF A NATIONAL RESEARCH AND PUBLIC POLICY AGENDA

Given the concerns about homeland security, confidentiality of proprietary information, and fraud in general, biometric recognition is becoming a routine method of recognizing individuals. If there is a pressing public policy need for which biometric systems are the most appropriate solution, understanding the science and technology issues is critical. As the preceding discussions should make clear, many questions remain.

The committee believes that more research into performance and robustness is needed. The lack of well-defined operational best practices based on solid science may allow governments and private organizations to issue overly vague or unrealistic mandates for biometric programs leading to poorly targeted oversight, delayed and troubled programs, excessive costs due to under- or overspecification of requirements, and failed deployments.

In short, the scientific basis of biometrics should be strengthened. Basic research should be done on the stability and distinctiveness of biometric traits; the control of environmental noise when acquiring samples; the correlation of biometric traits with private information, including medical conditions; and the demographic variability of biometric traits. Many fields of inquiry are relevant, even integral, to deepening the science of biometric recognition, including sensor design, signal processing, pattern recognition, human factors, statistics and biostatistics, computer systems design, information security, operations research, economics, politics, applied psychology, sociology, education, and the law.

Biometric systems perform well in many existing applications, but biometric capabilities and limitations are not yet well understood in very large scale applications involving tens of millions of users. Questions remain about whether today's biometric systems are sufficiently robust, able to handle errors when the consequences are severe. Although fingerprinting technology has been applied on a large scale for decades in law enforcement, human experts are available in this application to help process noisy or difficult samples. Even so, there have been a few high-profile misidentifications with serious ramifications. It remains to be seen if fully automatic biometric systems can meet performance requirements as the number and scale of deployments increase.

As mentioned above, a scientific basis is needed for the distinctiveness and stability of various biometric traits under a variety of collection processes and environments and across a wide population over decades. How accurately can a biometric trait be measured in a realistic operating environment? The individuality of biometric traits, their long- and short-term physiological and pathological variability, and their relationship to the providing population's genetic makeup, health, and other private attributes all merit research attention, which will require extensive data collection. The privacy protections to be afforded participants in such data collection need to be clearly outlined.

Improvements to biometric sensors and to the quality of the data acquired are crucial to minimizing recognition errors. Sensors should be made usable by a wider range of individuals in more environments and should be able to capture more faithfully (that is, with higher resolutions and with lower noise) underlying biometric traits of more than one kind in adverse situations and at a distance. Because many applications involve large numbers of sensors, attention should be paid to the development of low-cost but high-quality sensors. Additional areas meriting attention include representation and storage improvements and match-algorithm improvements.

Understanding how users interact with systems also merits further attention. The characteristics of the subject population, their attitudes and level of cooperation, the deployment environment, and procedures for measuring performance can all affect the system. Consequently, observation and experimentation in operational systems are required to understand how well biometric applications satisfy their requirements. Because of the challenges inherent in closely observing individuals, with or without their cooperation, human factors are critical to the design of processes for monitoring subjects and operators when assessing the effectiveness of a biometric system.

Another area where research is required is in the systems' view of biometric recognition, encompassing social, legal, and cultural aspects. Related are social implications of biometric recognition on a large scale. Research is needed, too, on the distinctive information security problems of biometric systems, such as defense against attacks by individuals using fake or previously captured biometric samples and the concealment of biometric traits, and on the protection of biometric reference databases. Decision analysis and threat modeling are other critical areas requiring research advances.

The U.S. government has created or funded several interdisciplinary, academically based research programs that provide a foundation for future work. Research support should aim for greater involvement of scientists and practitioners from relevant disciplines in biometric research, and studies should be published in the open, peer-reviewed scientific literature, with their stringently deidentified biometric samples made widely available to other researchers. A clearinghouse would facilitate efforts toward identifying standards implementation and interoperability issues, characterizing common elements of successful implementations, cataloging lessons learned, and maintaining data as input for testing product robustness and system performance.

Principle: As biometric recognition is deployed in systems of national importance, additional research is needed at virtually all levels of the system (including sensors, data management, human factors, and testing). The research should look at a range of questions from the distinctiveness of biometric traits to optimal ways of evaluating and maintaining large systems over many years.

1

Introduction and
Fundamental Concepts

From a very young age, most humans recognize each other easily. A familiar voice, face, or manner of moving helps to identify members of the family—a mother, father, or other caregiver—and can give us comfort, comradeship, and safety. When we find ourselves among strangers, when we fail to recognize the individuals around us, we are more prone to caution and concern about our safety.

This human faculty of recognizing others is not foolproof. We can be misled by similarities in appearance or manners of dress—a mimic may convince us we are listening to a well-known celebrity, and casual acquaintances may be incapable of detecting differences between identical twins. Nonetheless, although this mechanism can sometimes lead to error, it remains a way for members of small communities to identify one another.

As we seek to recognize individuals as members of larger communities, however, or to recognize them at a scale and speed that could dull our perceptions, we need to find ways to automate such recognition. Biometrics is the automated recognition of individuals based on their behavioral and biological characteristics.[1]

---

[1]"Biometrics" today carries two meanings, both in wide use. (See Box 1.1 and Box 1.2.) The subject of the current report—the automatic recognition of individuals based on biological and behavioral traits—is one meaning, apparently dating from the early 1980s. However, in biology, agriculture, medicine, public health, demography, actuarial science, and fields related to these, biometrics, biometry, and biostatistics refer almost synonymously to statistical and mathematical methods for analyzing data in the biological sciences. The two usages of

BOX 1.1
History of the Field—Two Biometrics

"Biometrics" has two meanings, both in wide use. The subject of this report—the automatic recognition of individuals based on biological and behavioral traits—is one meaning, which apparently dates from the early 1980s. In biology, agriculture, medicine, public health, demography, actuarial science, and fields related to these, "biometrics," "biometry," and "biostatistics" refer almost synonymously to statistical and mathematical methods for analyzing data in the biological sciences. This usage stems from the definition of biometry, proffered by the founder of the then-new journal *Biometrika* in its 1901 debut issue: "the application to biology of the modern methods of statistics." The writer was the British geneticist Francis Galton, who made important contributions to fingerprinting as a tool for identification of criminals, to face recognition, and to the central statistical concepts of regression analysis, correlation analysis, and goodness of fit.

Thus, the two meanings of "biometrics" overlap both in subject matter—human biological characteristics—and in historical lineage. Stigler (2000) notes that others had preceded the *Biometrika* founders in combining derivatives of the Greek βίος (bios) and μετρον (metron) to have specific meanings.[1] These earlier usages do not survive.

Johns Hopkins University opened its Department of Biometry and Vital Statistics (since renamed the Department of Biostatistics) in 1918. Graduate degree programs, divisions, and service courses with names incorporating "biostatistics," "biometrics," or "biometry" have proliferated in academic departments of health science since the 1950s. The American Statistical Association's 24 subject-matter sections began with the Biometrics Section in 1938, which in 1945 started the journal *Biometrics Bulletin*, renamed *Biometrics* in 1947. In 1950 *Biometrics* was transferred to the Biometric Society (now the International Biometric Society), founded in 1947 at Woods Hole, Massachusetts. The journal promotes "statistical and mathematical theory and methods in the biosciences through . . . application to new and ongoing subject-matter challenges." Concerned that *Biometrics* was overly associated with medicine and epidemiology, in 1996 the Society and the American Statistical Association jointly founded the *Journal of Agricultural, Biological, and Environmental Statistics* (*JABES*). The latter, along with other journals such as *Statistics in Medicine* and *Biostatistics*, have taken over the original mission of *Biometrika*, now more oriented to theoretical statistics.

Automated human recognition began with semiautomated speaker recognition systems in the 1940s. Semiautomated and fully automated fingerprint, handwriting, and facial recognition systems emerged in the 1960s as digital computers became more widespread and capable. Fully automated systems based on hand geometry

---

[1]S.M. Stigler, The problematic unity of biometrics, *Biometrics* 56: 653-658 (2000).

and fingerprinting were first deployed commercially in the 1970s, almost immediately leading to concerns over spoofing and privacy. Larger pilot projects for banking and government applications became popular in the 1980s. By the 1990s, the fully automated systems for both government and commercial applications used many different technologies, including iris and face recognition.

Clearly both meanings of biometrics are well-established and appropriate and will persist for some time. However, in distinguishing our topic from biometrics in its biostatistical sense, one must note the curiosity that two fields so linked in Galton's work should a century later have few points of contact. Galton wished to reveal the human manifestations of his cousin Charles Darwin's theories by classifying and quantifying personal characteristics. He collected 8,000 fingerprint sets, published three books on fingerprinting in four years,[2] and proposed the Galton fingerprint classification system extended in India by Azizul Haque for Edward Henry, Inspector General of Police, in Bengal. It was documented in Henry's book *Classification and Uses of Finger Prints.* Scotland Yard adopted this classification scheme in 1901 and still uses it.

But not all of Galton's legacy is positive. He believed that physical appearances could indicate criminal propensity and coined the term "eugenics," which was later used to horrific ends by the Third Reich. Many note that governments have not always used biologically derived data on humans for positive ends.

Galton's work was for understanding biological data. And yet biostatisticians, who have addressed many challenges in the fast-moving biosciences, have been little involved in biometric recognition research. And while very sophisticated statistical methods are used for the signal analysis and pattern recognition aspects of biometric technology, the systems and population sampling issues that affect performance in practice may not be fully appreciated. That fields once related are now separate may reflect that biometric recognition is scientifically less basic than other areas of interest, or that funding for open research is lacking, or even that most universities have no ongoing research in biometric recognition. A historical separation between scientifically based empirical methods developed specifically in a forensic context and similar methods more widely vetted in the open scientific community has been noted in other contexts and may also play a role here.[3,4]

---

[2]F. Galton, *Fingerprints* (1892); *Decipherment of Blurred Finger Prints* (1893); and *Fingerprint Directories* (1895). All were published by Macmillan in London.

[3]National Research Council, *The Polygraph and Lie Detection* (2003). Washington, D.C.: The National Academies Press, and National Research Council, *Strengthening Forensic Science in the United States: A Path Forward* (2009), Washington, D.C.: The National Academies Press.

[4]For more on the history of the field and related topics, see F. Galton, *On Personal Description*, Dublin, Ireland: Medical Press and Circular (1888), and S.J. Gould, *The Mis-measure of Man*, New York: Norton (1981).

BOX 1.2
A Further Note on the Definition of Biometrics

The committee defines biometrics as the automated recognition of individuals based on their behavioral and biological characteristics. This definition is consistent with that adopted by the U.S. government's Biometric Consortium in 1995. "Recognition" does not connote absolute certainty. The biometric systems that the committee considers always recognize with some level of error.

This report is concerned only with the recognition of human individuals, although the above definition could include automated systems for the recognition of animals. The definition used here avoids the perennial philosophical debate over the differences between "persons" and "bodies."[1] For human biometrics, an individual can only be a "body". In essence, when applied to humans, biometric systems are automated methods for recognizing bodies using their biological and behavioral characteristics. The word "individual" in the definition also limits biometrics to recognizing single bodies, not group characteristics (either normal or pathological). Biometrics as defined in this report is therefore not the tool of a demographer or a medical diagnostician nor is biometrics as defined here applicable to deception detection or analysis of human intent.

The use of the conjunction "and" in the phrase "biological and behavioral characteristics" acknowledges that biometrics is about recognizing individuals from observations that draw on biology and behaviors. The characteristics observable by a sensing apparatus will depend on current and, to the extent that the body records them, previous activities (for example, scars, illness aftereffects, physical symptoms of drug use, and so on).

---

[1]R. Martin and J. Barresi, *Personal Identity*, Malden, Mass.: Blackwell Publishing (2003); L.R. Baker, *Persons and Bodies: A Constitution View*, Cambridge, England: Cambridge University Press (2000).

Many traits that lend themselves to automated recognition have been studied, including the face, voice, fingerprint, and iris. A key characteristic of our definition of biometrics is the use of "automatic," which implies, at least here, that digital computers have been used.[2] Computers, in turn, require instructions for executing pattern recognition algorithms on trait samples received from sensors. Because biometric systems use sensed traits to recognize individuals, privacy, legal, and sociological factors are

---

"biometrics" overlap both in subject matter—human biological characteristics—and in historical lineage. This report's definition of biometrics is consistent with ISO/IEC JTC 1/SC 37 Standing Document 2, "Harmonized Biometric Vocabulary, version 10," August 20, 2008.

[2]Early biometric systems using analog computers and contemporary biometric systems using optical comparisons are examples of nondigital processing of biometric characteristics.

involved in all applications. Biometrics in this sense sits at the intersection of biological, behavioral, social, legal, statistical, mathematical, and computer sciences as well as sensor physics and philosophy. It is no wonder that this complex set of technologies called biometrics has fascinated the government and the public for decades.

The FBI's Integrated Automatic Fingerprint Identification System (IAFIS) and smaller local, state, and regional criminal fingerprinting systems have been a tremendous success, leading to the arrest and conviction of thousands of criminals and keeping known criminals from positions of trust in, say, teaching. Biometrics-based access control systems have been in continuous, successful use for three decades at the University of Georgia and have been used tens of thousands of times daily for more than 10 years at San Francisco International Airport and Walt Disney World.

There are challenges, however. For nearly 50 years, the promise of biometrics has outpaced the application of the technology. Many have been attracted to the field, only to leave as companies go bankrupt. In 1981, a writer in the *New York Times* noted that "while long on ideas, the business has been short on profits."[3] The statement continues to be true nearly three decades later. Technology advances promised that biometrics could solve a plethora of problems, including the enhancement of security, and led to growth in availability of commercial biometric systems. While some of these systems can be effective for the problem they are designed to solve, they often have unforeseen operational limitations. Government attempts to apply biometrics to border crossing, driver licenses, and social services have met with both success and failure. The reason for failure and the limitations of systems are varied and mostly ill understood. Indeed, systematic examinations that provide lessons learned from failed systems would undoubtedly be of value, but such an undertaking was beyond the scope of this report. Even a cursory look at such systems shows that multiple factors affect whether a biometric system achieves its goals. The next section, on the systems perspective, makes this point.

THE SYSTEMS PERSPECTIVE

One underpinning of this report is a systems perspective. No biometric technology, whether aimed at increasing security, improving throughput, lowering cost, improving convenience, or the like, can in and of itself achieve an application goal. Even the simplest, most automated, accurate, and isolated biometric application is embedded in a larger system. That system may involve other technologies, environmental factors, appeal policies shaped by security, business, and political considerations, or

---

[3]A. Pollack, Technology: Recognizing the real you, *New York Times*, September 9, 1981.

idiosyncratic appeal mechanisms, which in turn can reinforce or vitiate the performance of any biometric system.

Complex systems have numerous sources of uncertainty and variability. Consider a fingerprint scanner embedded in a system aimed at protecting access to a laptop computer. In this comparatively simple case, the ability to achieve the fingerprint scan's security objective depends not only on the biometric technology, but also on the robustness of the computing hardware to mechanical failures and on multiple decisions by manufacturer and employer about when and how the biometric technology can be bypassed, which all together contribute to the systems context for the biometric technology.

Most biometric implementations are far more complex. Typically, the biometric component is embedded in a larger system that includes environmental and other operational factors that may affect performance of the biometric component; adjudication mechanisms, usually at multiple levels, for contested decisions; a policy context that influences parameters (for example, acceptable combinations of cost, throughput, and false match rate) under which the core biometric technology operates; and protections against direct threats to either bypass or compromise the integrity of the core or of the adjudication mechanisms. Moreover, the effectiveness of such implementations relies on a data management system that ensures the enrolled biometric is linked from the outset to the nonphysical aspects of the enrolling individual's information (such as name and allowed privileges). The rest of this report should be read keeping in mind that biometric systems and technologies must be understood and examined within a systems context.

MOTIVATIONS FOR USING BIOMETRIC SYSTEMS

A primary motivation for using biometrics is to easily and repeatedly recognize an individual so as to enable an automated action based on that recognition.[4] The reasons for wanting to automatically recognize individuals can vary a great deal; they include reducing error rates and improving accuracy, reducing fraud and opportunities for circumvention, reducing costs, improving scalability, increasing physical safety, and improving convenience. Often some combination of these will apply. For example, almost all benefit and entitlement programs that have utilized

---

[4]Note that here we are using "recognition" colloquially—the biometrics community often uses this term as part of the sample processing task; it uses "verification" to mean that a sample matches a reference for a claimed identity and "identification" to mean the searching of a biometric database for a matching reference and the return of information about that individual.

biometrics have done so to reduce costs and fraud rates, but at the same time convenience may have been improved as well. See Box 1.3 for more on the variety of biometric applications.

Historically, personal identification numbers (PINs), passwords, names, social security numbers, and tokens (cards, keys, passports, and other physical objects) have been used to recognize an individual or to verify that a person is known to a system and may access its services or benefits. For example, access to an automatic teller machine (ATM) is generally controlled by requiring presentation of an ATM card and its corresponding PIN. Sometimes, however, recognition can lead to the denial of a benefit. This could happen if an individual tries to make a duplicate claim for a benefit or if an individual on a watch list tries to enter a controlled environment.

But reflection shows that authorizing or restricting someone because he or she knows a password or possesses a token is just a proxy for verifying that person's presence. A password can be shared indiscriminately or a physical token can be given away or lost. Thus, while a system can be

---

BOX 1.3
The Variety of Biometric Applications

Biometric technology is put to use because it can link a "person" to his or her claims of recognition and authorization within a particular application. Moreover, automating the recognition process based on biological and behavioral traits can make it more economical and efficient. Other motivations for automating the mechanisms for recognizing individuals using biometric systems vary depending on the application and the context in which the system is deployed; they include reducing error rates and improving accuracy; reducing fraud and circumvention; reducing costs; improving security and safety; improving convenience; and improving scalability and practicability. Numerous applications employ biometrics for one or more of these reasons, including border control and criminal justice (such as prisoner handling and process), regulatory compliance applications (such as monitoring who has access to certain records or other types of audits), determining who should be entitled to physical or logical access to resources, and benefits and entitlement management. The scope and scale of applications can vary a great deal—biometric systems that permit access might be used to protect resources as disparate as a nuclear power plant or a local gym. Even though at some level of abstraction the same motivation exists, the systems are likely to be very different and to merit different sorts of analysis, testing, and evaluation (see Chapter 2 for more on how application parameters can vary). The upshot of this wide variety of reasons for using biometric systems is that much more information is needed to assess the appropriateness of a given system for a given purpose beyond the fact that it employs biometric technology.

confident that the right password or token has been presented for access to a sensitive service, it cannot be sure that the item has been presented by the correct individual. Proxy mechanisms are even more problematic for exclusion systems such as watch lists, as there is little or no motivation for the subject to present the correct information or token if doing so would have adverse consequences. Biometrics offers the prospect of closely linking recognition to a given individual.

HUMAN IDENTITY AND BIOMETRICS

Essential to the above definition of biometrics is that, unlike the definition sometimes used in the biometrics technical community, it does not necessarily link biometrics to human identity, human identification, or human identity verification. Rather, it measures similarity, not identity. Specifically, a biometric system compares encountered biological/behavioral characteristics to one or more previously recorded references. Measures found to be suitably similar are considered to have come from the same individual, allowing the individual to be recognized as someone previously known to the system. A biometric system establishes a probabilistic assessment of a match indicating that a subject at hand is the same subject from whom the reference was stored.

If an individual is recognized, then previously granted authorizations can once again be granted. If we consider this record of attributes to constitute a personal "identity," as defined in the NRC report on authentication,[5] then biometric characteristics can be said to point to this identity record. However, the mere fact that attributes are associated with a biometric reference provides no guarantee that the attributes are correct and apply to the individual who provided the biometric reference.

Further, as there is no requirement that the identity record contain a name or other social identifier, biometric approaches can be used in anonymous applications. More concisely, such approaches can allow for anonymous identification or for verification of an anonymous identity. This has important positive implications for the use of biometrics in privacy-sensitive applications. However, if the same biometric measure is used as a pointer to multiple identity records for the same individual across different systems, the possibility of linking these records (and hence the various social identities of the same person) raises privacy concerns. See Box 1.4 for a note on privacy.

---

[5]National Research Council, *Who Goes There? Authentication Through the Lens of Privacy*, Washington, D.C.: The National Academies Press (2003).

BOX 1.4
A Note on Privacy

Privacy is an important consideration in biometric systems. The report *Who Goes There? Authentication Through the Lens of Privacy*,[1] focused on the intersection of privacy and authentication systems, including biometrics. Much of that analysis remains relevant to current biometric systems, and this report does not have much to add on privacy other than exploring some of the social and cultural implications of biometric systems (see Chapter 4). This reliance on an earlier report does not suggest that privacy is unimportant. Rather, the committee believes that no system can be effective without considerable attention to the social and cultural context within which it is embedded. The 2003 NRC report just referred to and its 2002 predecessor, which examined nationwide identity systems,[2] should be viewed as companions to this report.

---

[1]National Research Council, *Who Goes There? Authentication Through the Lens of Privacy*. Washington, D.C.: The National Academies Press (2003).
[2]National Research Council, *IDs—Not That Easy: Questions About Nationwide Identity Systems*. Washington, D.C.: The National Academies Press (2002).

The Fundamental Dogma of Biometrics

The finding that an encountered biometric characteristic is similar to a stored reference does not guarantee an inference of individualization—that is, that a single individual can be unerringly selected out of a group of all known individuals (or, conversely, that no such individual is known). The inference that similarity leads to individualization rests on a theory that one might call the fundamental dogma of biometrics:

An individual is more similar to him- or herself over time than to anyone else at any time.

This is clearly false in general; many singular attributes are shared by large numbers of individuals, and many attributes change significantly over an individual's lifetime. Further, it will never be possible to prove (or falsify) this assertion precisely as stated because "anyone else" will include all persons known or unknown, and we cannot possibly prove the assertion for those who are unknown.[6] In practice, however, we can relate similarity to individualization in situations where:

---

[6]The committee is aware of the Duhem/Quine and Popperian objections to provability in general of scientific theories.

An individual is more likely similar to him- or herself over time than to anyone else likely to be encountered.

This condition, if met, allows us to individualize through similarity, but with only a limited degree of confidence, based on knowledge of probabilities of encounters with particular biometric attributes. The goal in the development and applications of biometric systems is to find characteristics that are stable and distinctive given the likelihood of encounters. If they can be found, then the above conditions are satisfied and we have a chance of making biometrics work—to an acceptable degree of certainty—to achieve individualization.

A better fundamental understanding of the distinctiveness of human individuals would help in converting the fundamental dogma of biometrics into grounded scientific principles. Such an understanding would incorporate learning from biometric technologies and systems, population statistics, forensic science, statistical techniques, systems analysis, algorithm development, process metrics, and a variety of methodological approaches. However, the distinctiveness of biometric characteristics used in biometric systems is not well understood at scales approaching the entire human population, which hampers predicting the behavior of very large scale biometric systems.

The development of a science of human individual distinctiveness is essential to the effective and appropriate use of biometrics as a means of human recognition and encompasses a range of fields. This report focuses on the biometric technologies themselves and on the behavioral and biological phenomena on which they are based. These phenomena have fundamental statistical properties, distinctiveness, and varying stabilities under natural physiological conditions and environmental challenges, many aspects of which are not well understood.

BASIC OPERATIONAL CONCEPTS

In this section, the committee outlines some of the concepts underlying the typical operation of biometric systems in order to provide a framework for understanding the analysis and discussion in the rest of the report.[7] Two concepts are discussed: sources of (1) variability and

---

[7]There have been several comprehensive examinations of biometrics technologies and systems over the years. See, for example, J.L. Wayman, A.K. Jain, D. Maltoni, and D. Maio, eds., *Biometric Systems: Technology, Design, and Performance Evaluation*, London: Springer (2005); J. Woodward, Jr., N. Orlans, and P. Higgins, *Biometrics: Identity Assurance in the Information Age*, New York: McGraw-Hill/Osborne Media (2002); and A.K. Jain, R. Bolle, and S. Pankanti, eds., *Biometrics: Personal Identification in a Networked Society*, Norwell, Mass.: Kluwer Academic Press (1999). The National Science and Technology Council also recently

(2) uncertainty in biometric systems and modalities, including multibiometric approaches.

Sample Operational Process

The operational process typical for a biometric system is given in Figure 1.1. The main components of the system for the purposes of this discussion are the capture (whereby the sensor collects biometric data from the subject to be recognized), the reference database (where previously enrolled subjects' biometric data are held), the matcher (which compares presented data to reference data in order to make a recognition decision), and the action (whereby the system recognition decision is revealed and actions are undertaken based on that decision.[8]

This diagram presents a very simplified view of the overall system. The operational efficacy of a biometric system depends not only on its technical components—the biometric sample capture devices (sensors) and the mathematical algorithms that create and compare references—but also on the end-to-end application design, the environment in which the biometric sensor operates, and any conditions that impact the behavior of the data subjects, that is, persons with the potential to be sensed.

For example, the configuration of the database used to store references against which presented data will be compared affects system performance. At a coarse level, whether the database is networked or local is a primary factor in performance. Networked databases need secure communication, availability, and remote access privileges, and they also raise more privacy challenges than do local databases. Local databases, by contrast, may mean replicating the reference database multiple times, raising security, consistency, and scalability challenges.[9] In both cases, the accuracy and currency of any identification data associated with reference

---

issued reports that elaborate on biometrics systems with an eye to meeting government needs. See, for example, "The National Biometrics Challenge," available at http://www. biometrics.gov/Documents/biochallengedoc.pdf, and "NSTC Policy for Enabling the Development, Adoption and Use of Biometric Standards," available at http://www.biometrics. gov/Standards/NSTC_Policy_Bio_Standards.pdf.

[8]The data capture portion of the process has the most impact on accuracy and throughput and has perhaps been the least researched portion of the system. The capture process, which involves human actions (even in covert applications) in the presence of a sensor, is not well understood. While we may understand sensor characteristics quite well, the interaction of the subject with the sensors merits further attention. See Chapter 5 for more on research opportunities in biometrics.

[9]Both the NRC report *Who Goes There?* (2003) and the EC Data Protection Working Party discuss the implications of centralized or networked data repositories versus local storage of data. The latter is available at http://ec.europa.eu/justice_home/fsj/privacy/docs/wpdocs/2003/wp80_en.pdf.

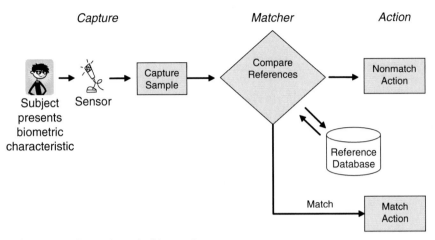

FIGURE 1.1 Operation of a biometric system.

characteristics in a biometric system are independent of the likelihood that a sample came from the individual who provided the reference. In other words, increasing confidence in a recognition result does not commensurately increase our confidence in the validity of any associated data.

Measures of Operational Efficacy

Key aspects of operational efficacy include recognition error rates; speed; cost of acquisition, operation, and maintenance; data security and privacy; usability; and user acceptance. Generally, trade-offs must be made across all of these measures to achieve the best-performing system consistent with operational and budgetary needs. For example, recognition error rates might be improved by using a better but more time-consuming enrollment process; however, the time added to the enrollment process could result in queues (with loss of user acceptance) and unacceptable costs.

In this report the committee usually discusses recognition error rates in terms of the false match rate (FMR; the probability that the matcher recognizes an individual as a different enrolled subject) and the false nonmatch rate (FNMR; the probability that the matcher does not recognize a previously enrolled subject). FMR and FNMR refer to errors in the matching process and are closely related to the more frequently reported false acceptance rate (FAR) and the false rejection rate (FRR). FAR and FRR refer to results at a broader system level and include failures arising from additional factors, such as the inability to acquire a sample. The committee uses these terms less frequently as they can sometimes introduce confusion between the semantics of "acceptance" and "rejection" in terms of the claimed performance for biometric recognition versus that for the overall application. For example, in a positive recognition system, a false acceptance occurs when subjects are recognized who should not be recognized—either because they are not enrolled in the system or they are someone other than the subject being claimed. In this case the sense of false acceptance is aligned for both the biometric matching operation and the application function. In a system designed to detect and prevent multiple enrollments of a single person, sometimes referred to as a negative recognition system, a false acceptance results when the system fails to match the submitted biometric sample to a reference already in the database. If the system falsely matches a submitted biometric sample to a reference from a different person, the false match results in a denial of access to system resources (a false rejection).

Variability and Uncertainty

Variability in the biometric data submitted for comparison to the enrolled reference data can affect performance. As mentioned above, the matching algorithm plays a role in how this variability is handled. However, many other factors can influence performance, depending on how the specific biometric system is implemented. Generally speaking, biometric applications automatically capture aspects of one or more human traits to produce a signal from which an individual can be recognized. This signal cannot be assumed to be a completely accurate representation of the underlying biometric characteristic. Biometric systems designers and experts have accepted for some time that noise in the signal occurs haphazardly; while it can never be fully controlled, it can be modeled probabilistically.[10] Also, not all uncertainty about biometric systems is due to random noise. Uncertainty pervades a biometric system in a number of ways. Several potential sources of uncertainty or variation are discussed here and should be kept in mind when reading the rest of this report. These sources are listed based on the order of events that take place as an individual works with the system to gain recognition.

• Depending on the biometric modality, information content presented by the same subject at different encounters may be affected by changes in age, environment, stress, occupational factors, training and prompting, intentional alterations, sociocultural aspects of the situation in which the presentation occurs, changes in human interface with the bio-

---

[10]See for example, J.P. Campbell, Jr., *Testing with the YOHO CD-ROM Voice Verification Corpus*, The Biometrics Consortium, available at http://www.biometrics.org/REPORTS/ICASSP95.html.

metric system, and so on. These factors are important both at the enroll-ment phase and during regular operation. The next section describes within-person and between-person variability in depth.

• Sensor operation is another source of variability. Sensor age and calibration can be factors, as well as how precisely the system-human interface at any given time stabilizes extraneous factors. Sensitivity of sensor performance to variation in the ambient environment (such as light levels) can play a role.

• The above sources may be expected to induce greater variation in the information captured by different biometric acquisition systems than in the information captured by the same system. Other factors held constant, information on a single subject captured from repeated encoun-ters with the same sensor will vary less than that captured from different sensors in the same system, which will vary less than that captured from encounters with different systems.

In addition to information capture, performance variation across bio-metric systems and when interfacing components of different systems depends on how the information is used, including the following:

• Differences in feature extraction algorithms affect performance, with effects sometimes aggravated by the need for proprietary systems to be interoperable.

• Differences between matching algorithms and comparison scor-ing mechanisms. How these algorithms and mechanisms interact with the preceding sources of variability of information acquired and features extracted also contributes to variation in performance of different systems. For instance, matching algorithms may differ in their sensitivity to biologi-cal and behavioral instability of the biometric characteristic over time, as well as the characteristic's susceptibility to intentional modification.

• The potential for individuals attempting to thwart recognition for one reason or another is another source of uncertainty that systems should be robust against. See below for a more detailed discussion on security for biometric systems.

In light of all of this, determining an appropriate action to take—where possible actions include to recognize, to not recognize, or to transition the system to a secondary recognition mechanism based on the signal from a biometric device—involves decision making under uncertainty.

Within- and Between-Person Variability

Variability in the observed values of a biometric trait can refer to variation in a given trait observed in the same person or to variation in the trait observed in different persons. Effective overall system performance requires that within-person variability be small—the smaller the better—relative to between-person variability.

Within-Person Variation

Ideally, every time we measure the biometric trait of an individual, we should observe the same patterns. In practice, however, the different samples produce different patterns, which result in different digital representations (references). Such within-person variation, sometimes referred to as "intraclass variation," typically occurs when an individual interacts differently with the sensor (for example, by smiling) or when the biometric details of a person (for example, hand shape and geometry or iris pigmentation details) change over time. The sensing environment (for example, ambient lighting for a face and background noise for a voice) can also introduce within-person variation. There are a number of ways to reduce or accommodate such variation, including controlled acquisition of the data, storage of many references for every user, and systematic updating of references. Reference updating, although essential to any biometric system since it can help account for changes in characteristics over time, introduces system vulnerabilities. Some biometric traits are more likely to change over time than others. Face, hand, and voice characteristics, in particular, can benefit from suitably implemented reference update mechanisms. Within-person variation can also be caused by behavioral changes over time.

Between-Person Variation

Between-person variation, sometimes referred to as "interclass variation," refers generally to person-to-person variability. Since there is an inherent similarity between biometric traits among some individuals (faces of identical twins offer the most striking example), between-person variation between two individuals may be quite small. Also, a chosen digital representation (the features) for a particular biometric trait may not very effective in separating the observed patterns of particular subjects. In contrast, demographic heterogeneity among enrolled subjects in a biometric system database may contribute to large between-person variation in measurements of a particular biometric trait, although fluctuations in the sensing environment from which their presentation samples are obtained may contribute to large within-person variation as well.

It is the magnitude of within-person variation relative to between-person variation (observed in the context of a finite range of expression of human biometric traits) that determines the overlap between distributions of biometric measurements from different individuals and hence limits the number of individuals that can be discriminated and recognized by a biometric system with acceptable accuracy. When within-person variation is small relative to between-person variation, large biometric systems with high accuracy are feasible because the distributions of observed biometric data from different individuals are likely to remain widely separated, even for large groups. When within-person variation is high relative to between-person variation, however, the distributions are more likely to impinge on each other, limiting the capacity of a recognition system. In other words, the number of enrolled subjects cannot be arbitrarily increased for a fixed set of features and matching algorithms without degrading accuracy. When considering the anticipated scale of a biometric system, the relative magnitudes of both within-person and between-person variations should be kept in mind.

Stability and Distinctiveness at Global Scale

Questions of variation and uncertainty become challenging at scale. In particular, no biometric characteristic, including DNA, is known to be capable of reliably correct individualization over the size of the world's population. Factors that make unlikely the discovery of a characteristic suitably stable and distinctive over such a large population include the following:

• *Individuals without traits.* Almost any trait that can be noninvasively observed will fail to be exhibited by some members of the human population, because they are missing the body part that carries the trait, because environmental or occupational exposure has eradicated or degraded the trait, or because their individual expression of the trait is anomalous in a way that confuses biometric systems.

• *Similar individuals.* In sufficiently small populations it is highly likely that almost any chosen trait will be sufficiently distinctive to distinguish individuals. As populations get larger, most traits (and especially most traits that can be noninvasively observed) may have too few variants to guarantee that different individuals are distinguishable from one another. The population statistics for most biometric traits are poorly understood.

• *Feature extraction effects.* Even in cases where a biometric trait is distinctive, the process of converting the analog physical property of a human to a digital representation that can be compared against the properties of other individuals involves loss of detail—that is, information loss—and introduction of noise, both of which can obscure distinctions between individuals.

However, the lack of an entirely stable and distinctive characteristic at scale need not stand in the way of effective use of biometrics if the system is well designed. Some biometric systems might not have to deal with all people at all times but might need only to deal with smaller groups of people over shorter periods of time. It may be possible to find traits that are sufficiently stable and distinctive to make many types of applications practicable.

Implicit in all biometric systems are estimated probabilities of the sameness and difference of source of samples and stored references, separately for presenters of different types, such as residents and impostors. This leads to the explicit use of ratios of probabilities in some biometric recognition algorithms. Because assessing the likelihood that a sample came from any particular reference may involve computing similarity to many references through use of ratios of probabilities or other normalization techniques, it cannot be strictly said that any form of recognition involves comparison of only one sample to one known reference. However, verification of a claim of similarity of a sample with a specific reference may appear to those unfamiliar with the algorithmic options to involve only a single comparison. This form of verification is often referred to as "one to one." Verification of claims of similarity to an unspecific reference is often referred to as a "one-to-many" application because many comparisons to assess similarity over the enrolled individuals are required.

Biometric Modalities

A biometric modality[11] refers to a system built to recognize a particular biometric trait. Face, fingerprint, hand geometry, palm print, iris, voice, signature, gait, and keystroke dynamics are examples of biometric traits.[12] In the context of a given system and application, the presentation of a user's biometric feature involves both biological and behavioral aspects. Some common biometric modalities described by Jain et al. (2004)[13] are

---

[11]A biometric modality is the combination of a biometric trait, sensor type, and algorithms for extracting and processing the digital representations of the trait. When any two of these three constituents differ from one system to the next, the systems are said to have different modalities. For example, infrared facial recognition and iris recognition are different modalities since the trait and the algorithms differ even if the same camera is used.

[12]DNA could be considered a biometric modality if the technologies for it can be sufficiently automated. However, this report focuses on those modalities and systems for which automated technologies are further along in development and deployment.

[13]Anil K. Jain, Arun Ross, and Salil Prabhakar, An introduction to biometric recognition, *IEEE Transactions on Circuits and Systems for Video Technology*, Special Issue on Image- and Video-Based Biometrics 14(1) (2004).

summarized briefly here. Many of the issues associated with biometric systems (and correspondingly much of the discussion in this report) are not modality-specific, although of course the choice of modality has implications for system design and, potentially, system performance.

Face

Static or video images of a face can be used to facilitate recognition. Modern approaches are only indirectly based on the location, shape, and spatial relationships of facial landmarks such as eyes, nose, lips, and chin, and so on. Signal processing techniques based on localized filter responses on the image have largely replaced earlier techniques based on representing the face as a weighted combination of a set of canonical faces. Recognition can be quite good if canonical poses and simple backgrounds are employed, but changes in illumination and angle create challenges. The time that elapses between enrollment in a system and when recognition is attempted can also be a challenge, because facial appearance changes over time.

Fingerprints

Fingerprints—the patterns of ridges and valleys on the "friction ridge" surfaces of fingers—have been used in forensic applications for over a century. Friction ridges are formed in utero during fetal development, and even identical twins do not have the same fingerprints. The recognition performance of currently available fingerprint-based recognition systems using prints from multiple fingers is quite good. One factor in recognition accuracy is whether a single print is used or whether multiple or ten-prints (one from each finger) are used. Multiple prints provide additional information that can be valuable in very large scale systems. Challenges include the fact that large-scale fingerprint recognition systems are computationally intensive, particularly when trying to find a match among millions of references.

Hand Geometry

Hand geometry refers to the shape of the human hand, size of the palm, and the lengths and widths of the fingers. Advantages to this modality are that it is comparatively simple and easy to use. However, because it is not clear how distinctive hand geometry is in large populations, such systems are typically used for verification rather than identification. Moreover, because the capture devices need to be at least the size of a hand, they are too large for devices like laptop computers.

Palm Print

Palm prints combine some of the features of fingerprints and hand geometry. Human palms contain ridges and valleys, like fingerprints, but are much larger, necessitating larger image capture or scanning hardware. Palm prints, like fingerprints, have particular application in the forensic community, as latent palm prints can often be found at crime scenes.

Iris

The iris, the circular colored membrane surrounding the eye's pupil, is complex enough to be useful for recognition. The performance of systems using this modality is promising. Although early systems required significant user cooperation, more modern systems are increasingly user friendly. However, although systems based on the iris have quite good FMRs, the FNMRs can be high. Further, the iris is thought to change over time, but variability over a lifetime has not been well characterized.[14]

Voice

Voice directly combines biological and behavioral characteristics. The sound an individual makes when speaking is based on physical aspects of the body (mouth, nose, lips, vocal cords, and so on) and can be affected by age, emotional state, native language, and medical conditions. The quality of the recording device and ambient noise also influence recognition rates.

Signature

How a person signs his or her name typically changes over time. It can also be strongly influenced by context, including physical conditions and the emotional state of the signer. Extensive experience has also shown that signatures are relatively easy to forge. Nevertheless, signatures have been accepted as a method of recognition for a long time.

Gait

Gait, the manner in which a person walks, has potential for human recognition at a distance and potentially, over an extended period of

---

[14]See Sarah Baker, Kevin W. Bowyer, and Patrick J. Flynn, Empirical evidence for correct iris match score degradation with increased time lapse between gallery and probe images, *International Conference on Biometrics*, pp. 1170-1179 (2009). Available at http://www.nd.edu/~kwb/BakerBowyerFlynnICB_2009.pdf.

time. Laboratory gait recognition systems are based on image processing to detect the human silhouette and associated spatiotemporal attributes. Gait can be affected by several factors, including choice of footwear, the walking surface, and clothing. Gait recognition systems are still in the development stage.

Keystroke

Keystroke dynamics are a biometric trait that some hypothesize may be distinctive to individuals. Indeed, there is a long tradition of recognizing Morse code operators by their "fists"—the distinctive patterns individuals used to create messages. However, keystroke dynamics are strongly affected by context, such as the person's emotional state, his or her posture, type of keyboard, and so on.

Comparison of Modalities

Each biometric modality has its pros and cons, some of which were mentioned in the descriptions above. Moreover, even if some of the downsides could be overcome, a modality itself might have inherent deficiencies, although very little research into this has been done. Therefore, the choice of a biometric trait for a particular application depends on issues besides the matching performance. Raphael and Young identified a number of factors that make a physical or a behavioral trait suitable for a biometric application.[15] The following seven factors are taken from an article by Jain et al.:[16]

- *Universality.* Every individual accessing the application should possess the trait.
- *Uniqueness.* The given trait should be sufficiently different across members of the population.
- *Permanence.*[17] The biometric trait of an individual should be sufficiently invariant over time with respect to a given matching algorithm. A trait that changes significantly is not a useful biometric.
- *Measurability.* It should be possible to acquire and digitize the biometric trait using suitable devices that do not unduly inconvenience the

[15]D.E. Raphael and J.R. Young, *Automated Personal Identification*, Palo Alto, Calif.: SRI International (1974).

[16]Jain A.K., Bolle R., and Pankanti, S., *Biometrics: Personal Identification in Networked Society*, Norwell, Mass.: Kluwer Academic Publisher (1999).

[17]In this report the committee generally refers to the stability of a trait rather than its permanence.

individual. Furthermore, the acquired raw data should be amenable to processing to extract representative features.

• *Performance.* The recognition accuracy and the resources required to achieve that accuracy should meet the requirements of the application.

• *Acceptability.* Individuals in the target population that will use the application should be willing to present their biometric trait to the system.

• *Circumvention.* The ease with which a biometric trait can be imitated using artifacts—for example, fake fingers in the case of physical traits and mimicry in the case of behavioral traits—should conform to the security needs of the application.

Multibiometrics

As the preceding discussions make clear, using a single biometric modality may not always provide the performance[18] needed from a given system. One approach to improving performance (error rates but not speed) is the use of multibiometrics, which has several meanings:[19]

• *Multisensors.* Here, a single modality is used, but multiple sensors are used to capture the data. For example, a facial recognition system might employ multiple cameras to capture different angles on a face.

• *Multiple algorithms.* The same capture data are processed using different algorithms. For example, a single fingerprint can be processed using minutiae and texture. This approach saves on sensor and associated hardware costs, but adds computational complexity.

• *Multiple instances.* Multiple instances of the same modality are used. For example, multiple fingerprints may be matched instead of just one, as may the irises of both eyes. Depending on how the capture was done, such systems may or may not require additional hardware and sensor devices.

• *Multisamples.* Multiple samples of the same trait are acquired. For example, multiple angles of a face or multiple images of different portions of the same fingerprint are captured.

• *Multimodal.* Data from different modalities are combined, such as face and fingerprint, or iris and voice. Such systems require both hard-

---

[18]The term "performance," used in the biometrics community generally, refers broadly to error rates, processing speed, and data subject throughput. See, for example, http://www.biometrics.gov/Documents/Glossary.pdf.

[19]A. Ross and A.K. Jain, Multimodal biometrics: An overview. *Proceedings of 12th European Signal Processing Conference.* Available online at http://biometrics.cse.msu.edu/Publications/Multibiometrics/RossJain_MultimodalOverview_EUSIPCO04.pdf.

ware (sensors) and software (algorithms) to capture and process each modality being used.

Hybrid systems that combine the above also may prove useful. For example, one could use multiple algorithms for each modality of a multimodal system. The engineering of multibiometric systems presents challenges as does their evaluation. There are issues related to the architecture and operation of multibiometrics systems and questions about how best to model such systems and then use the model to drive operational aspects. Understanding statistical dependencies is also important when using multibiometrics. For example, Are the modalities of hand geometry and fingerprints completely independent—beyond, say, the trivial correlation between a missing hand and the failure to acquire fingerprints? As a large-scale biometric system becomes multimodal, it is that much more important to adopt approaches and architectures that support interoperability and implementation of best-of-breed matching components. This would mean, for example, including matching software, image segmentation software, and sample quality assessment software as they become available.[20] This approach to interoperability support would also facilitate replacing an outdated matcher with a newer, higher performing matcher without having to scrap the entire system and start from scratch. Likewise, new multibiometric fusion algorithms could be implemented without requiring a major system redesign. Finally, new human interface issues may come into play if multiple observations are needed of a single modality or of multiple modalities.[21]

COPING WITH THE PROBABILISTIC
NATURE OF BIOMETRIC SYSTEMS

The probabilistic aspect of biometric systems is often missing from popular discussions of the technology. For the purposes of this discussion, the committee will ignore FAR and FRR and consider only the

---

[20]For example, this would allow for the implementation of two different 10-print matchers, one of which might be better for the rapid processing of high-quality fingerprints and the other of which might be better for poor-quality fingerprints but take longer to process.

[21]Historically, multibiometric systems have proved more expensive, time consuming, and difficult to implement than single modal systems. These drawbacks are blamed by the principals for the nondeployment of the Department of Defense (DOD) Base and Installation Security System (BISS) in the 1970s (A. Fejfar and J. Myers, *The Testing of 3 Automatic ID Verification Techniques for Entry Control*, 2nd International Conference on Crime Countermeasures, Oxford, England, July 25-29, 1977, and A. Fejfar, *Combining Techniques to Improve Security in Automatic Access Control*, Carnahan Conference on Crime Countermeasures, University of Kentucky, May 17-19, 1978).

sources of error that contribute to the false match rates (FMR) and false nonmatch rates (FNMR). In this context, the FMR is the probability that the incorrect trait is falsely recognized and the FNMR is the probability that a correct trait is falsely not recognized. The probability a correct trait is truly recognized is 1 − FNMR and the probability an incorrect trait is truly not recognized is 1 − FMR. Complicating matters, biometric match probabilities are only one part of what we need to help predict the real-world performance of biometric systems.[22]

It seems intuitively obvious that a declared nonmatch in a biometric system with both FMRs and FNMRs of 0.1 percent is almost certainly correct. Unfortunately, intuition is grossly misleading in this instance, and the common misconception can have profound sociological impacts (for example, it might lead to the assumption that a suspected criminal is guilty if the fingerprints or DNA samples from the suspect "match" those at the crime scene). Understanding why this natural belief is often wrong is one of the keys to understanding how to use biometrics effectively. From the perspective of statistical decision theory, it is not enough to focus on error rates. All they provide is the conditional probability of a recognition/nonrecognition given that the presenting individual should be recognized and the conditional probability of a nonrecognition given that the presenting individual should not be recognized.

To illustrate, we will consider first an access control system. Let us assume that we know by experience or experiment the probability that a false claim by an impostor will be accepted and the probability that a true claim by a legitimate user will be accepted. In an operational environment, however, we would like to know the converse: the probability that the claimant is an impostor, given that the claim was accepted by the system. That is, we wish to know the probability of a false claim given a recognition by the system. To perform such a probability inversion, it is necessary to use Bayes' theorem. A fundamental characteristic of this theorem is that it requires the prior probability of a false claim (impostor) (see Box 1.5). That is, some information about the frequency of false claims/impostors is needed in order to know the probability that any given recognition by the system is in error. The following series of examples illustrates how the percentage of "right" decisions by a biometric system depends upon the impostor base rate,[23] the percentage of "impostors" actually encountered by the system, not just on the error rates of the technology. The error rates

---

[22]In addition, if recognition rates are tunable in a given system (that is, if it is possible to adjust certain parameters and make, say, the FMR or FNMR higher or lower), that has implications for system architecture.

[23]The "imposter base rate" refers to the probability that a randomly chosen individual presenting him- or herself to the biometric system will be an impostor.

BOX 1.5
Decision Theory Components Required for Biometric Recognition

A formal decision-theoretic formulation of biometric recognition, whether from a classical frequentist or a Bayesian perspective, is beyond the scope of this report. However, the flavor of such formulations is conveyed by some of their components:

- *State space.* A mutually exclusive and exhaustive listing of the uncertain components relevant to decisions about biometric recognition. This may be as simple as whether or not a presenting individual is an authorized user or an impostor.
- *Action space.* A mutually exclusive and exhaustive listing of the decisions to be made or actions to be taken based on the result of the biometric matcher. These might simply be to declare a match or a nonmatch (regardless of whether this decision is or is not the correct one). Slightly more complex action spaces would indicate the level of authorization of a presenting individual or whether the presenting individual should be required to have further screening.
- *Probability function.* The probability distribution of the comparison score for each of the elements (impostor/authorized) in the state space.
- *Prior distribution.* The probability of the elements (impostor/authorized) of the state space. The prior distribution should accurately reflect the probability of the state of the presenting individual based on all information prior to obtaining the comparison score from the biometric device.
- *Consequences.* The cost or utility of each (state, action) combination.

Here is a simple formulation of Bayes' theorem in this context:

P (Impostor | Biometric Match) =

$$\frac{P(Biometric\ Match\ |\ Impostor)\ P(Impostor)}{P(Biometric\ Match\ |\ Impostor)\ P(Impostor) + P(Biometric\ Match\ |\ Not\ Impostor)\ P(Not\ Impostor)}$$

In this formulation, $P(B|A)$ is the conditional probability of B given A, that is, the fraction of instances that B is true among instances when A is true, and the prior probability of a false claim is $P(Impostor)$. Also, $P(Not\ Impostor) = 1 - P(Impostor)$.

(the FMR and FNMR) are independent of the impostor base rate, but all of these pieces of information are needed to understand the frequency that a given recognition (or nonrecognition) by the system is in error.

To return to the example above, imagine that we have installed a rather accurate biometric verification system to control entry to a college dormitory. Suppose that the system has a 0.1 percent FMR and a 0.1 percent FNMR. The system lets an individual into the dorm if it matches the individual to a stored biometric reference—if the system does not find a match, it does not let the individual in. We would like to know how often a nonmatch represents an attempt by a nonresident "impostor" to get into the dorm.[24] The answer, it turns out, is "it depends."[25]

First, consider the case where the impostor base rate is 0 percent—that is, no impostors ever try to get into the dorm. In this case, all of the people using the biometric system are residents. Since the system has a 0.1 percent FNMR, it will generate a false nonmatch once every 1,000 authentication attempts. All of these nonmatches will be errors (because in this case all the people using the system are residents). In this case, over time we will discover that our confidence in a nonmatch is zero—because nonmatches are always false. Table 1.1 contains the calculations for this case. Figure 1.2 presents the information in Table 1.1.

Now consider a different case: For every 999 times a resident attempts entry, one nonresident impostor tries to get into the building. In this case, since the system has a 0.1 percent FMR, it will (just as in the preceding case) generate one false nonmatch for each 1,000 recognition attempts. But since the system also has a 0.1 percent FNMR, it will (with 99.9 percent probability) generate a nonmatch for the one nonresident impostor. On the average, therefore, every 1,000 recognition attempts will include one impostor (who will generate a correct nonmatch with overwhelming probability) and one resident who will generate an incorrect nonmatch. There will therefore be two nonmatches—50 percent of them correct and 50 percent of them incorrect—in every 1,000 authentication attempts. In this case, using the same system as in the preceding case, with the same sensor and the same 0.1 percent FNMR, we will observe 50 percent true nonmatches and 50 percent false nonmatches. Table 1.2 shows how to calculate confidence that a nonmatch will be true in this case. Figure 1.3 presents the information in Table 1.2.

Tables 1.3 and 1.4 calculate confidence in the truth of a nonmatch in cases where the impostor base rate is 1 percent (that is, where 1 percent of the people trying to get into the dorm are nonresident impostors) and in cases where the impostor base rate is 50 percent (that is, half the people trying to get into the dorm are nonresident impostors). (Figures 1.4 and 1.5 present calculations of the data in Tables 1.3 and 1.4, respectively.) Note that confidence in the truth of a nonmatch approaches 99.9 percent

---

[24]For the purposes of the discussion, we assume that the imposter is ready and able to enter the dorm and in possession of any information or tokens needed to initiate the bio-metric verification process.

[25]This discussion and associated examples draw heavily on *Weight-of-Evidence for Forensic DNA Profiles* by David J. Balding (New York: Wiley, 2005) and are based primarily on material in its Chapter 2, "Crime on an Island," and Chapter 3, "Assessing Evidence via Likelihood Ratios."

TABLE 1.1 Impostor Base Rate of 0%

Proffered Identity	Authentication Attempts	Biometric Decision		Conclusion
		Match	Nonmatch	
Authentic	1,000	1,000 × 99.9% = 999	1,000 × 0.1% = 1	Confidence that a nonmatcher is an impostor = fraction of impostors among nonmatches = 0/1 = 0%
Impostor	0	0 × 0.1% = 0	0 × 99.9% = 0	
Total	1,000	999	1	

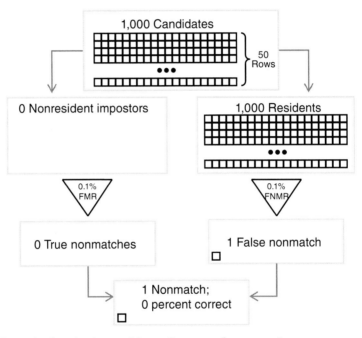

FIGURE 1.2 Authenticating residents (impostor base rate 0 percent; very low nonmatch accuracy).

(the true nonmatch rate of the system) only when at least half the people trying to get into the dorm are impostors!

These examples teach two lessons:

1. It is impossible to specify accurately the respective fractions of a biometric system's matches and nonmatches that are correct without knowing how many individuals who "should" match and how many individuals who "should not" match are presenting to the system.

2. A biometric technology's FMR and its FNMR are not accurate mea-

TABLE 1.2 Impostor Base Rate of 0.1%

Proffered Identity	Authentication Attempts	Biometric Decision		Conclusion
		Match	Nonmatch	
Authentic	999	999 × 99.9% = 998	999 × 0.1% = 1	Confidence that a nonmatcher is an impostor = fraction of impostors among nonmatches = 1/2 = 50%
Impostor	1	1 × 0.1% = 0	1 × 99.9% = 1	
Total	1,000	998	2	

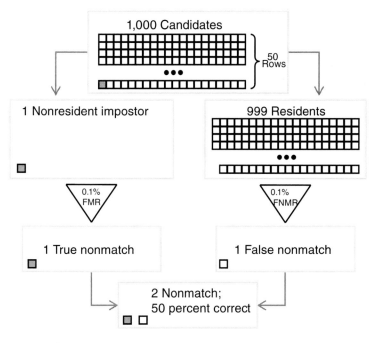

FIGURE 1.3 Authenticating residents (impostor base rate 0.1 percent; moderate nonmatch accuracy).

sures of how often the system gives the right answer in an operational environment and will in many cases greatly overstate the confidence we should have in the system.

The bad news, therefore, is that even with a very accurate biometric system, correctly identifying rare events (an impostor's attempt to get into the dorm, in our first example) is very hard. The good news, by the same token, is that if impostors are very rare, our confidence in correctly identifying people who are not impostors (that is, determining that we are

TABLE 1.3 Impostor Base Rate of 1.0%

Proffered Identity	Authentication Attempts	Biometric Decision		Conclusion
		Match	Nonmatch	
Authentic	990	990 × 99.9% = 989	990 × 0.1% = 1	Confidence that a nonmatcher is an impostor = 10/11 = 91%
Impostor	10	10 × 0.1%= 0	10 × 99.9% = 10	
Total	1,000	989	11	

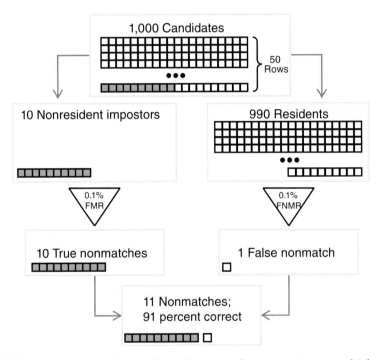

FIGURE 1.4 Authenticating residents (impostor base rate 1 percent; high non-match accuracy).

not in a rare-event situation) can be very high—far higher than 99.9 percent. In our first example, when the impostor base rate is 0.1 percent, our confidence in the correctness of a match is almost 100 percent (actually 99.9999 percent)—much higher than suggested by the FMR and FNMR. It is easy to see why this is true. Almost everyone who approaches the sensor in the dorm is actually a resident. For residents, all of whom are supposed to match, false matches are possible (one resident could claim to

TABLE 1.4 Impostor Base Rate of 50%

Proffered Identity	Authentication Attempts	Biometric Decision		Conclusion
		Match	Nonmatch	
Authentic	500	500 × 99.9% = 499	500 × 0.1% = 1	Confidence that a nonmatcher is an impostor = 499/500 = 99.8%
Impostor	500	500 × 0.1% = 1	500×99.9% = 499	
Total	1,000	500	500	

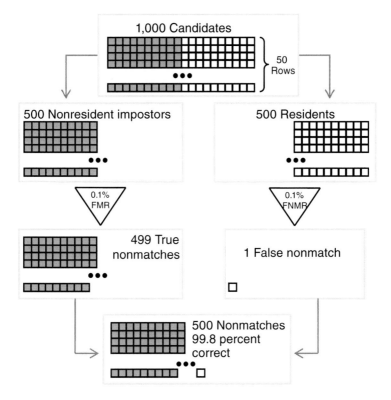

FIGURE 1.5 Authenticating residents (impostor base rate 50 percent; very high nonmatch accuracy).

be another resident and match with that other resident's reference), but a false match never results in a false acceptance, since a false match has the same system-level result—entrance to the dorm—as correct identification. A false match is possible only when an impostor approaches the sensor and is incorrectly matched. But almost no impostors ever approach the

TABLE 1.5 Impostor Base Rate of 0.1%

Proffered Identity	Authentication Attempts	Biometric Decision		Conclusion
		Match	Nonmatch	
Authentic	999	999 × 99.9% = 998	999 × 0.1% = 1	Confidence that a matcher is not an impostor = fraction of nonimpostors among matches = 998/998 = 100%
Impostor	1	1 × 0.1% = 0	1 × 99.9% = 1	
Total	1,000	998	2	

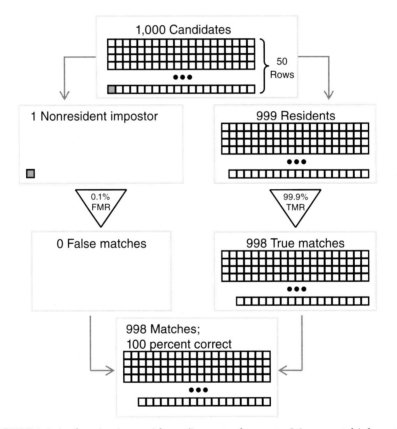

FIGURE 1.6 Authenticating residents (impostor base rate 0.1 percent; high match accuracy).

door, and (because the technology is very accurate) impostors who do approach the door are very rarely incorrectly matched. Table 1.5 provides the information for this case, and Figure 1.6 illustrates the case. Note that Figure 1.6 depicts only matches, in contrast to Figures 1.2 through 1.5, which depict only nonmatches.

The overall lesson is that as the impostor base rate declines in a recognition system, we become more confident that a match is correct but less confident that a nonmatch is correct. Examples of this phenomenon are common and well documented in medicine and public health. People at very low risk of a disease, for example, are usually not routinely screened, because positive results are much more likely to be a false alarm than lead to an early diagnosis. Unless the effects of the base rate on system performance are anticipated in the design of a biometric system, false alarms may consume large amounts of resources in situations where very few impostors exist in the system's target population. Even more insidiously, frequent false alarms could lead screeners to a more lax attitude (the problem of "crying wolf") when confronted with nonmatches. Depending on the application, becoming inured to nonmatches could be costly or even dangerous.[26]

ADDITIONAL IMPLICATIONS FOR OPEN-SET IDENTIFICATION SYSTEMS

The above discussion concerned an access control application. Large-scale biometric applications may be used for identification to prevent fraud arising from an individual's duplicate registration in a benefits program or to check an individual's sample against a "watch list"—a set of enrolled references of persons to be denied benefits (such as passage at international borders). This is an example of an open-set identification system, where rather than verifying an individual's claim to an identity, the system determines if the individual is enrolled and may frequently be processing individuals who are not enrolled. The implications of Bayes' theorem are more difficult to ascertain in this situation because here biometric processing requires comparing a presenting biometric against not just a single claimed enrollment sample, but against unprioritized enrollment samples of multiple individuals. Here, as above, the chances of erroneous matches and erroneous nonmatches still depend on the frequencies with which previously enrolled and unenrolled subjects present to the system. Such chances also depend on the length of the watch list and on how this length and the distribution of presenters[27] to the system

---

[26]These costs vary with the application. For instance, if Walt Disney World is able to prevent most people who do not pay from getting into its theme park, then erroneously admitting a few who only pretended to pay may not be all that important. The importance of an imposter on an airplane may be much greater. So the assessment of uncertainty has to take into account the importance of certainty for different outcomes.

[27]The ratio of presenters who are enrolled subjects on the watch list to presenters who are not.

interrelate. The overall situation is complex and requires detailed analysis, but some simple points can be made.

In general, additions to a watch list offer new opportunities for an unenrolled presenter to match with the list, and for an enrolled presenter to match with the wrong enrollee. If additions to the watch list are made in such a way as to leave the presentation distribution unchanged—for example, by enrolling persons who will not contribute to the presentation pool—then the ratio of true to false matches will decline, necessarily reducing confidence in a match. Appendix B formalizes this argument, incorporating a prior distribution for the unknown proportion of presenters who are previously enrolled.

We may draw an important lesson from this simple situation: Increasing list size cannot be expected to improve all aspects of system performance. Indeed, in an identification system with a stable presentation distribution, as list length increases we should become less confident that a match is correct.

A comment from the Department of Justice Office of the Inspector General's report on the Mayfield mistake[28] exemplifies this point: "The enormous size of the FBI IAFIS [Integrated Automated Fingerprint Identification System] database and the power of the IAFIS program can find a confusingly similar candidate print. The Mayfield case indicates the need for particular care in conducting latent fingerprint examinations involving IAFIS candidates because of the elevated danger in encountering a close nonmatch."[29]

But this is not the end of the story, because in some circumstances changes in watch-list length may be expected to alter the presentation distribution. The literature distinguishes between open-set identification systems, in which presenters are presumed to include some persons not previously enrolled, and closed-set identification systems, in which presentations are restricted to prior enrollees. Closed-set identification systems meet the stable presentation distribution criterion de jure, so that the baseline performance response still automatically applies to the expanded list. But the actual effect of list expansion on system performance, when the presentation distribution in an open-set identification system may change, will depend on the net impact of modified per-presenter error rates and the associated rebalancing of the presentation distribution. In other words, the fact that the list has expanded may affect who is part

---

[28]Brandon Mayfield, an Oregon attorney, was arrested by the FBI in connection with the Madrid train bombings of 2004 after a fingerprint on a bag of detonators was mistakenly identified as belonging to Mayfield.

[29]See http://www.usdoj.gov/oig/special/s0601/PDF_list.htm; http://www.usdoj.gov/oig/special/s0601/exec.pdf.

of the pool of presenters. This rebalancing may occur without individuals changing their behavior simply because of the altered relationship between the length of the watch list and the size of the presenting population. But it may also occur as a result of intentional behavior change by new enrollees, who may stop or reduce their presentations to the system as a response to enrollment. Clearly, increasing watch list size without very careful thought may decrease the probability that an apparent matching presenter is actually on the list.

That lengthening a watch list may reduce confidence in a match speaks against the promiscuous searching of large databases for individuals with low probability of being on the list, and it tells us that we must be extremely careful when we increase "interoperability" between databases without control over whether increasing the size of the list has an impact on the probability that the search subject is on the list. Our response to the apparent detection of a person on a list should be tempered by the size of the list that was searched. These lessons contradict common practice.

The designers of a biometric system face a challenge: to design an effective system, they must have an idea of the base rate of detection targets in the population that will be served by the system. But the base rate of targets in a real-world system may be hard to estimate, and once the system is deployed the base rate may change because of the reaction of the system's potential detection targets, who in many cases will not want to be detected by the system. To avoid detection, potential targets may avoid the system entirely, or they may do things to confuse the system or force it into failure or backup modes in order to escape detection. For all these reasons, it is very difficult for the designers of the biometric system to estimate the detection target base rate accurately. Furthermore, no amount of laboratory testing can help to determine the base rate. Threat modeling can assist in developing estimates of imposter base rates and is discussed in the next section.

SECURITY AND THREAT MODELING

Security considerations are critical to the design of any recognition system, and biometric systems are no exception. When biometric systems are used as part of authentication applications, a security failure can lead to granting inappropriate access or to denying access to a legitimate user. When biometric systems are used in conjunction with a watch list application, a security failure can allow a target of investigation to pass unnoticed or cause an innocent bystander to be subjected to inconvenience, expense, damaged reputation, or the like. In seeking to understand the security of biometric systems, two security-relevant processes are of interest: (1) the determination that an observed trait belongs to a living human who is present and is acting intentionally and (2) the proper matching (or non-matching) of the observed trait to the reference data maintained in the system.

Conventional security analysis of component design and system integration involves developing a threat model and analyzing potential vulnerabilities—that is, where one might attack the system. As described above, any assessment of the effectiveness of a biometric system (including security) requires some sense of the impostor base rate. To estimate the impostor base rate, one should develop a threat model appropriate to the setting.[30] Biometric systems are often deployed in contexts meant to provide some form of security, and any system aimed at security requires a well-considered threat model.[31] Before deploying any such system, especially on a large scale, it is important to have a realistic threat model that articulates expected attacks on the system along with what sorts of resources attackers are likely to be able to apply. Of course, a thorough security analysis, however, is not a guarantee that a system is safe from attack or misuse. Threat modeling is difficult. Results often depend on the security expertise of the individuals doing the modeling, but the absence of such analysis often leads to weak systems.

As in all systems, it is important to consider the potential for a malicious actor to subvert proper operation of the system. Examples of such subversion include modifications to sensors, causing fraudulent data to be introduced; attacks on the computing systems at the client or matching engine, causing improper operation; attacks on communication paths between clients and the matching engine; or attacks on the database that alter the biometric or nonbiometric data associated with a sample.

A key element of threat modeling in this context is an understanding of the motivations and capabilities of three classes of users: clients, imposters, and identity concealers. Clients are those who should be recognized by the biometric system. Impostors are those who should not be recognized but will attempt to be recognized anyway. Identity concealers are those who should be recognized but are attempting to evade recognition. Important in understanding motivation is to envision oneself as the

[30]For one discussion of threat models, see Microsoft Corporation, "Threat Modeling" available at http://msdn.microsoft.com/en-us/security/aa570411.aspx. See also Chapter 4 in NRC, *Who Goes There? Authentication Through the Lens of Privacy* (2003).

[31]The need to consider threat models in a full system context is not new nor is it unique to biometrics. In his 1997 essay "Why Cryptography Is Harder Than It Looks," available at http://www.schneier.com/essay-037.html, Bruce Schneier addresses the need for clearly understanding threats from a broad perspective as part of integrating cryptographic components in a system. Schneier's book *Secrets and Lies: Digital Security in a Networked World* (New York: Wiley, 2000) also examined threat modeling and risk assessment. Both are accessible starting points for understanding the need for threat modeling.

impostor or identity concealer.[32] Some of the subversive population may be motivated by malice toward the host (call the malice-driven subversive data subject an attacker), others may be driven by curiosity or a desire to save or gain money or time, and still others may present essentially by accident. This mix would presumably depend on characteristics of the application domain:

- The value to the subversive subject of the asset claimed—contrast admission to a theme park and physical access to a restricted research laboratory.
- The value to the holder of the asset to which an attacker claims access—say, attackers intent on vandalism.
- The ready accessibility of the biometric device.
- How subversive subjects feel about their claim being denied or about detection, apprehension, and punishment.

A threat model should try to answer the following questions:

- What are the various types of subversive data subjects?
- Is it the system or the data subject who initiates interaction with the biometric system?
- Is auxiliary information—for example, a photo ID or password—required in addition to the biometric input?
- Are there individuals who are exempt from the biometric screening—for example, children under ten or amputees?
- Are there human screening mechanisms, formal or informal, in addition to the automated biometric screening—for example, a human attendant who is trained to watch for unusual behavior?
- How can an attack tree[33] help to specify attack modes available to a well-informed subversive subject?
- Which mechanisms can be put in place to prevent or discourage repeated attempts by subversive subjects?

Here are some further considerations in evaluating possible actions to be taken:

---

[32]This is often referred to as a "red team" approach—see, for example, the description of the Information Design Assurance Red Team at Sandia National Laboratories, at http://idart.sandia.gov/.

[33]For a brief discussion of attack trees, see G. McGraw, "Risk Analysis: Attack Trees and Other Tricks," August 1, 2002. Available at http://www.drdobbs.com/184414879.

• Will the acceptance of a false claim seriously impact the host organization's mission or damage an important protected resource?

• Have all intangibles (for example, reputation, biometric system disruption) been considered?

• How would compromise of the system—for example, acceptance of a false claim for admission to a secure facility—damage privacy, release or degrade the integrity of proprietary information, or limit the availability of services?

In summary, as discussed at length above, FMRs and FNMRs in themselves are insufficient to describe or assess the operational performance of a biometric system and may be seriously misleading. It is necessary to also anticipate the fraction of reported matches that are likely to be true matches and the fraction of reported nonmatches that are likely to be true nonmatches. The analysis above shows that these will vary greatly with the base rate of impostors presenting to the system. The base rate should be estimated using one or more reasonable threat models. Biometric system design should then incorporate this information, as well as the costs and/or utilities of actions resulting from true and false matches and nonmatches.

All information systems are potentially vulnerable, but the design of biometric systems calls for special considerations:

• *Probabilistic recognition.* This fundamental property of biometric systems means that risk analysis must recognize that there is a probability of making incorrect recognitions (positive or negative). If an attacker can gain access to a large-scale biometric database, then he or she has the opportunity to search for someone who is a biometric doppelganger—someone for whom there is a close enough match given the target false match rate for the system.[34]

• *Exposure of biometric traits.* This can occur either through direct observation or through access to biometric databases. It allows an attacker to create fraudulent copies of those traits to be used in an attempt to mislead a biometric sensor.[35]

---

[34]See the discussion on biometric risk equations in T.E. Boult and R. Woodworth, Privacy and security enhancements in biometrics, *Advances in Biometrics: Sensors, Algorithms and Systems*, N.K. Ratha and V. Govindaraju, eds., New York: Springer (2007).

[35]For an example of how a fingerprint image can be transferred to a gelatinous material and then used to mislead a finger-imaging sensor, see T. Matsumoto, H. Matsumoto, K. Yamada, and S. Hoshino, Impact of artificial "gummy" fingers on fingerprint systems, *Optical Security and Counterfeit Deterrence Techniques IV*: 4677 (2002). For a discussion on improvements to techniques for reconstituting fingerprint images from processed template data, see J. Feng and A.K. Jain, FM model-based fingerprint reconstruction from minutiae template,

- *Concealment of biometric traits.* In some applications, such as those intended to prevent multiple enrollments or to identify persons on watch lists, an attacker can avoid detection by concealing biometric traits through relatively simple actions.[36]

These considerations show the importance of how a sample is presented. Since biometric data must be thought of as public information, a system must take appropriate precautions to verify that the sample presented belongs to the individual presenting the sample and that the individual is voluntarily presenting himself or herself for identification. In some cases, this may mean supervised recognition; in others it may mean that a technical mechanism is employed to validate the sensor and that the sensor can differentiate genuine samples from fraudulent synthesized samples. In the latter case, an appropriate sensor takes the place of a human observer of the presentation ceremony.

In response to growing concerns over identity theft and fraud, some advocates suggest that legislation be enacted to prohibit the selling or sharing of an individual's biometric data.[37] By making it illegal to traffic in biometric data and by requiring encryption for the storage of biometric data in authentication systems, the hope is that the chance of inappropriate data disclosure will be reduced significantly, preserving the utility of biometric authentication for widespread use. However, the encryption of data does not guarantee that the underlying data will not be exposed. Furthermore, covert observation of many biometric traits is possible, making acquisition of these data hard to avoid. Accordingly, the more ubiquitous biometric systems become, the more important it is that each system using biometrics perform a threat analysis that presumes public knowledge of a subject's biometric traits. Those systems should then deploy measures to verify that the presentation ceremony is commensurate with the risk of impersonation.[38] Furthermore, in high-assurance and high-criticality

---

in *Advances in Biometrics*, Massimo Tistarelli and Mark S. Nixon, eds., Third International Conference, Alghers, Italy (2009).

[36]Sometimes identity is concealed using more radical techniques. Recently it was reported that a Chinese national had had her fingerprints surgically transferred from one hand to another to avoid recognition by Japanese border control. Available at http://mdn.mainichi.jp/mdnnews/news/20091207p2a00m0na010000c.html. The Interpol fingerprint department provides a historical perspective on fingerprint alteration at http://www.interpol.int/Public/Forensic/fingerprints/research/alteredfingerprints.pdf.

[37]P. Swire, "Lessons for biometrics from SSNs and identity fraud," presentation to the committee on March 15, 2005.

[38]Note, for instance, the recent shutdown of the "Clear" air traveler program by Verified Identity Pass. The company held personal information, including biometric information, on thousands of individuals. At the time of this writing, a court had enjoined the company from selling the data.

applications, biometric recognition should not be the sole authentication factor.

ON REPORT SCOPE AND BOUNDARIES

This report explores the strengths and limitations of biometric systems and their legal, social, and philosophical implications. One core aim of the report is to dispel the common misconception that a biometric system unambiguously recognizes people by sensing and analyzing their biometric characteristics. No biometric technology is infallible; all are probabilistic and bring uncertainty to the association of an individual with a biometric reference, some of it related to the particular trait being scrutinized by the system. Variability in biometric traits also affects the probability of correct recognition. In the end, probability theory must be well understood and properly applied in order to use biometric systems effectively and to know whether they achieve what they promise.

This report does not address whether a biometric system is the best way to meet a particular application goal. It does not compare biometric technologies with potential alternatives for particular applications, because such alternatives would have to be evaluated case by case. This chapter has reviewed the fundamental properties of biometric systems. Chapter 2 will offer a framework for considering the requirements of an application from the engineering standpoint. Chapter 3 outlines lessons learned from other types of systems. Chapter 4 examines the social, cultural, and legal issues related to biometric systems. Finally, Chapter 5 summarizes the research challenges and open questions identified in the earlier chapters.

2

Engineering Biometric Systems

The preceding chapter described many of the fundamental concepts that underlie biometric systems. Of equal importance is the engineering of these systems. Moreover, while design, engineering and development of component parts of the systems are important, it is the development of a biometric system as a whole that is most critical to successful system deployment. The scope of a biometric system is broad and includes not only basic operations such as enrollment or matching but also user training and the adjudication process for dealing with contested results and exception handling in general. A holistic view that accounts for human interaction, and not simply the combination of sensors and matchers, is needed. Biometric systems are best considered in their deployment context, including all their particularities such as function, environment, and user population.

A systems engineering view is especially important when the systems are to be used on a large scale, such as for border control or social service entitlement, when all the best practices associated with system design and management are called for. While the evolution of sensor devices, matching technologies, and human factors can dominate the attention of system designers, the ultimate success of the overall system generally relies on attention to conventional system development issues. These include:

- A clear understanding of the system's functional objectives, including the user population and environmental constraints of the system's deployment context.

- A model for accommodating the evolution of technology over time.
- An understanding of the effectiveness of the system through frequent or continuous operational testing.
- A model for validating the financial value of the system over time.
- A strong understanding of the human interactions with the system and how they change with time.
- A holistic security model that addresses the application security context and provides a cogent analysis of the potential for fraudulent presentation of biometric samples.

Successful biometric applications require top-down conceptualization, with clear delineation of purpose within a systems context. Technology-driven implementations could inadvertently target a secondary rather than a primary objective and fail to anticipate mission-critical aspects of the context in which this human-centric technology is applied, including sources of challenges to the system. This chapter starts with an overview of biometric system operations. It then turns to the application requirements of biometric systems, including application parameters that can affect system performance. A brief discussion of test and evaluation concludes the chapter.

BASIC BIOMETRIC SYSTEM OPERATIONS

The operations performed by a generic biometric system are the capture and storage of enrollment (reference) biometric samples and the capture of new biometric samples and their comparison with corresponding reference samples. Figure 2.1 depicts the operation of a generic biometric system and is a more detailed version of the sample operational diagram from Chapter 1, although some systems will differ in their particulars.

Enrollment Operations

Enrollment of a new subject into a biometric system is achieved by performing the functions denoted in the upper processes A, B, C, and E of Figure 2.1. The samples are analyzed to ensure their quality is consistent with the matching algorithms to be used later for comparison with opera-

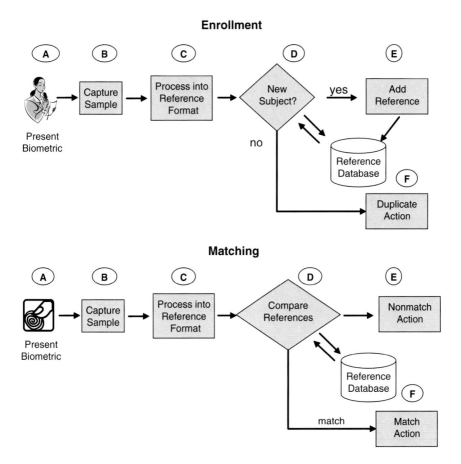

FIGURE 2.1 Idealized operations of a generic biometric system.

tionally obtained samples.[1] The data samples are then processed to form references that are stored for future comparison in a database or on transportable media such as a smart card.[2] Depending on the requirements of

---

[1]However, quality assessment algorithms for biometric enrollment are not universally available and may need to be specific to the proprietary feature extraction and comparison algorithms used later in the process. In some applications, such as use of facial recognition with images stored by passport issuance agencies on electronic passports, there will be no knowledge during the enrollment process of what algorithms might be used later to compare the stored image. This is true for visas as well.

[2]In other applications, such as facial recognition biometrics with images stored on passports, processing of the reference sample is deferred until the matching process is initiated.

the system, functions D and F might also be performed against previously stored references to detect attempts at duplicate enrollment. Ensuring that the reference format generated by function C conforms to pertinent standards such as those issued by ANSI or ISO[3] would aid data interoperability between systems or within an evolving system over time.

If sample quality is inadequate to create a reference likely to be correctly matched in future comparisons, the biometric characteristic might be resampled until adequate quality is obtained. However, the ability to resample to control quality of the enrollment sample will depend upon both system function and policy. If adequate quality is unattainable, then policy will dictate either that the best of the samples captured is retained or that the subject is declared a "failure to enroll." In the latter case, a fallback approach might be established for the individual. For example, in an employee access control system, the particular employee might be allowed entry with only a card and a PIN (two-factor authentication) or by requiring that a security guard verify the individual's facial photo. For enrollment in a watch list system, however, system management would have to balance the impact on future accuracy of retaining poor-quality references against having no reference for a person of interest. Such balancing has been a major challenge for systems used in forensic, military, and intelligence applications.

Enrollment of a subject also properly links the digital representation of the biometric reference with identity attributes established for use in this application.[4] Biometric traits acquired at later times are then recognized by comparing with the enrolling reference.[5] Hence, the quality and representativeness of the enrolling reference are crucial to later success, and variation in these can contribute to errors in the recognition process.

Most of the factors that affect the quality and representativeness of the enrollment reference involve the interaction of the subject with the collection device or irregularities in or absence of the biometric trait in

---

[3]See the Registry of U.S. Government Recommended Biometric Standards (Version 1.0, approved June 5, 2008) for a catalog of suitable standards for interoperability of U.S. federal biometric applications. Available at http://www.biometrics.gov/standards/Standards Registry.pdf.

[4]In the general case, no assumptions are made about the presumed "true identity" of the enrolling subject. The process of enrolling the subject's biometric data allows for subsequent retrieval of identity attributes. The legitimacy of the subject's association with those attributes, however, must be established by means outside the biometric enrollment process. See FIPS PUB 201-1 for an example of a process for verifying identity claims. Available at http://csrc.nist.gov/publications/fips/fips201-1/FIPS-201-1-chng1.pdf.

[5]Some systems may update enrolled references over time as a side effect of matching actions to minimize the drift of a biometric trait. In general, controls comparable to those applied at initial enrollment should be in place to minimize the opportunity to inject fraudulent data into the biometric database.

the subject, with some involving the device itself. Generally, the operational conditions under which both references and subsequent biometric samples are collected for a criminal justice system (such as booking a crime suspect) allow more control of quality than other types of systems. For image-based biometric modalities these may include the stability and positioning relative to a desired standard during imaging, the presence of distracting or camouflaging appurtenances, calibration and cleanliness of the sensing device, the image review process and criteria for certifying successful enrollment, and, indirectly, factors that affect the ability to control these. Such indirect but important factors may include the degree to which the acquisition of these biometric references is supervised and the staffing levels, throughput goals and pressure to achieve them, aspects of the environment that can affect sensor and human performance, the degree to which the subject is aware of and cooperative with the acquisition, and the efficiency of handling special exceptions, such as subjects who are physically or mentally challenged in their ability to present the required biometric or whose references are repeatedly rejected.

For some systems the reference representation is too large or complicated for efficient storage and high-throughput computer processing. An alternative is to extract mathematical abstractions known as "features" (or models of those features) from the reference samples and store only those, discarding the reference samples. Another alternative is to store reference samples using some standard compression technique (such as JPEG), not processing the sample to produce features until they are needed for comparison. In principle the choice of biometric features, the degree of independent information and hence distinctiveness conveyed by each additional feature, and the balance between feature multiplicity and storage efficiency can have major effects on recognition error rates. This is borne out by the large performance differences exhibited in tests that compare products using the same biometric trait.[6]

Storage of the reference, whether in the form of samples, features, or models, completes the enrollment process. Subsequently, samples provided by the enrollee and others claiming or claimed to be the enrollee will be compared against the reference. In some cases, for example, in the case of a law enforcement database and an enrollee who is the object of a security or law enforcement search, the reference may be added to a watch list containing many millions of samples that might ultimately be checked.

---

[6]For example, the NIST Minutiae Interoperability Exchange Test. See http://fingerprint.nist.gov/minex/.

Capture and Matching Operations

The processes A, B, C, D, E, and F in Figure 2.1 are carried out to capture one or more biometric samples and compare them against the reference(s). When possible, a sample to be used for comparison against a stored reference is analyzed to ensure adequate quality consistent with the feature extraction and matching algorithms used. [7] When the user is cooperative and the sample quality is inadequate, acquisition attempts may be repeated up to a permitted maximum number of times until a suitable sample is obtained. If adequate quality is not attained, depending on policy, either the best sample is used or "failure to acquire" is declared and a fallback procedure invoked, as noted earlier. Acceptable samples are converted to a format suitable for comparing against the stored references, generally following the same methods used to create the corresponding references.

If verification is the desired operation, the reference for the identity claimed is retrieved from the database (perhaps triggered by the user presenting a PIN or a proximity card), a comparison is made, and a comparison score is assigned. If this score exceeds a threshold, a match is declared. Otherwise a nonmatch is declared or possibly, for an intermediate window of scores, an indeterminate result is declared and a cooperative subject may be asked to resubmit the same biometric sample, an alternative biometric sample, or take additional action, such as contacting a security guard to execute a manual fallback procedure. If the desired operation is identification, then the sample features are compared against a portion or all of the reference database. In this case, one or more matches are indicated if the comparison score exceeds a predetermined threshold value. The top scoring match exceeding a threshold may be indicated, or all matches may be adjudicated by a human examiner.

This discussion has outlined the structure of a process that cannot be expected to work perfectly, deterministically providing the "right" answers, because it has inherent opportunities for error and uncertainty. It should be noted that the accuracy associated with declaring a match or nonmatch at a given score threshold is probabilistic and associated with metrics such as the false match rate (FMR) and false nonmatch rate (FNMR). So it is important to think carefully not only about what thresh-

---

[7]Adequate sample quality is a somewhat vague characterization of what is needed. In this context we are concerned with the utility of the sample in the biometric system. This utility can be maximized when a clearly expressed trait is captured with high fidelity. If the data subject's traits are poorly expressed owing to, for example, highly abraded finger friction ridges, then repeated attempts to capture a sample useful for comparison are unlikely to succeed.

olds are appropriate but also about all the various ways uncertainties can enter the system, as described in Chapter 1.

OPERATIONAL CONTEXT

Many factors affect the success or failure of a biometric system in its implementation. Table 2.1 illustrates the common parameters of the data subject, application, and technology contexts that affect both system design decisions and operational performance. Some parameter labels in Table 2.1 are shortened to fit and are then elaborated on below. Parameter intensities are directionally ordered left to right by increasing difficulty of implementation. For example, biometric systems of which participating subjects are aware and for which they are cooperatively motivated are easier to implement successfully than systems in which sample acquisition is surreptitious and subjects are hostile to the purpose. This list is incomplete but intended to spur the thinking of prospective system implementers. Each row is discussed below, grouped by user, application, technology, and performance contexts.

For any given class of applications or, more precisely, for any target deployment, one can begin an analysis of where on this table the demands

TABLE 2.1 Parameters That Affect System Design Decisions and System Effectiveness

Parameter	Degree or Intensity from High to Low		
User context			
Data subject awareness	Very		Not very
Data subject motivated	Very		Not very
Data subject well-trained	Very		Not very
Data subject habituated	Very		Not very
Who benefits?	Both	User/consumer	Owner/agency
Application context			
Application supervised	Very		Not very
Application type	Positive claim		Negative claim
Application type	Verification	One to few matching	Identification
Data interoperability	Closed	Supposed to be closed	Open
Technology context			
Environment controlled	Very		Not very
Passive versus active	Active	Passive w/cooperation	Passive
Covert versus overt	Overt		Covert
Performance context			
Throughput requirements	Low	Medium	High
Sensitivity to error rate requirements	Low	Medium	High

of the application lie. For example, consider a system to allow entrance to a gym, where data subject awareness and motivation would be high, the training of data subjects would be medium to low, the system probably requires active cooperation, the environment is controlled, and throughput is probably not a significant consideration. By contrast, a border control system using biometrics will be faced with users who are not well trained and perhaps not well motivated but will nonetheless have high throughput and stringent error rate requirements and so on. Clearly, stating that a system is a biometric system or uses "biometrics" does not provide much information about what the system is for or how difficult it is to successfully implement. Moreover, the parameters listed here are limited to factors that arise in day-to-day operations. There are broader system requirements (such as security, availability, and so on) that will also vary a great deal depending on the application and will also bear on the design, development, and, ultimately, operation of the system.

User Context

Data Subject Awareness

Does the data subject know that a biometric sample is being collected? Watch-list applications may use references gathered under different circumstances and from other systems, so even if the data subject is aware that a biometric characteristic is being collected, he or she might not know that the data are being collected for the purpose of biometric recognition. For example, a police station might take a mug shot photo when booking a suspect. Although the subject would be very aware that a photo was being taken, he or she might not know that the photo would later become part of a facial recognition watch list in an airport.

Data Subject Motivation

Does the data subject want to present the feature to the system in a way that is repeatable? Some subjects—such as in a typical physical or logical access control application where the subject is enrolled in the system, has valid rights, and wants access privileges—are motivated because matching is needed to perform a job function or to access to an entitlement, a privilege, or money. Other data subjects, such as prisoners, may be very unmotivated to interact with the system. This lack of motivation could be due to the adverse outcomes of a correct match or the desire to deliberately attempt to deceive the system by faulty interaction. Motivation and cooperation are closely aligned but not identical. A data subject can be cooperating with a system but still not be motivated to use it. In other cases, such as use of facial images on passports, the data subject may be motivated to present an image deemed "attractive" rather than one that is truly representative and repeatable.

Data Subject Training

Has the subject been instructed in the proper use of the system? Is there an opportunity in the enrollment process to provide feedback on correct or incorrect feature presentation? Many applications may not give data subjects a chance to have human interaction with system staff during enrollment or subsequent uses. Is there an opportunity to inform the subject before exposure to the system? Different system types may require more training than others for correct usage. A more general question is, What kind of training or operational cues or feedback to subjects will improve system usability and performance?

Data Subject Habituation

How often does the subject use the system? Does the subject use the system and get feedback on a regular basis for a long time (for example, long enough to develop muscle memory if any is required)?[8] Some systems that require active participation offer more assistance than others, guiding subjects to intuitively perform the functions necessary to give high-quality biometric measures. Infrequent usage, long gaps between uses, or poor ergonomic design can lead to poor positioning, which affects system accuracy and thereby throughput (see the discussion of technology context below).

Who Benefits?

Who receives the benefits of implementing a biometric recognition system in a particular application scenario? If the system is used in place of an existing system (for example, replacing verification using a driver's license), then both the deployer of the system and the data subjects may expect to obtain the same benefits as from the existing system. A system meant to offer convenience to users and save time for its deployers may also save time for the users and may be more convenient for the deployers. Both parties gain each other's intended benefits. In other situations

---

[8]A NIST study on the effects of habituation on fingerprint quality found that habituation in the absence of feedback failed to affect the quality of fingerprint collection. Habituation with feedback did lead to improved fingerprint quality over time from a cooperative population. Available at http://zing.ncsl.nist.gov/biousa/docs/WP302_Theofanos.pdf.

the benefits may be one-sided, especially where the biometric system is employed as an additional layer in an existing system. Take the example of adding a biometric system as a third factor to an existing card-plus-PIN physical access control system. A subject must spend additional time at a portal and faces an increased chance of being denied access because of the additional challenge and appears, moreover, to gain no direct benefit. The deployer of the system, however, substantially increases its confidence that the person entering is in fact the authorized user, with all other things being equal. So the need for security on the part of the system deployer is transferred to increased burden on the user. Even in a replacement implementation, this situation could exist. In a prison application, the use of a biometric system in lieu of an alternative system of prisoner identification offers no particular benefits from a prisoner perspective, and the prisoner is probably not interested in being verified as him or herself; however, the operators of the prison have a very high interest in proper prisoner identification relative to many other factors such as convenience and speed. Accuracy is the main benefit sought, but the data subject may have limited stake in this and little motivation to cooperate.

Application Context

Supervision

Is the system staffed? Is there someone present in the immediate vicinity who could give instructions or advice, help with questions, or solve problems with system usage, or is the system unstaffed? There may be a middle ground where there are persons nearby who can help but who are not immediately accessible or who have limited knowledge about the system and can help only to a certain degree. A system also may be self-service, with no opportunity for assistance other than, perhaps, a phone call.

Positive vs. Negative Claim Systems

Systems whose purpose is to verify a claim that a data subject is known to the system are considered "positive claim" systems. These systems often require a claim by the data subject to a specific reference and by extension to the corresponding enrollment record. However, alternative examples of positive claim systems, needing the unspecific claim "I am enrolled," have existed since the early 1990s. These systems were previously called "PINless verification" systems. Their function is to verify the claim without returning any information on which enrollment record ("who") the data subject corresponds to. Negative claim systems verify the claim that a data subject is not enrolled in the system. Examples of such systems include systems to detect fraudulent multiple enrollments in social service entitlement programs or driver's license registration systems. Note that some systems could be viewed and used as both positive and negative claim systems, but generally not simultaneously. A driver's license or an identification card system could verify a negative claim at the time of enrollment and then verify a positive claim later when the identification card is used to assert an identity. Some systems, such as watch lists, might on the surface appear to be ambiguous with regard to negative or positive claim—with the nature of the claim depending on whether the point of view is that of the data subject or the system administrators. However, such systems are nearly always classified from the point of view of the data subject, because his or her viewpoint will determine the nature of system vulnerabilities.

Verification vs. Identification Systems

Verification systems are used to check an individual's claim to be a subject enrolled in the system. These systems perform a so-called one-to-one type of match, usually by the data subject providing a name, number, token, or password that points to or unlocks the subject's enrollment reference.[9] Identification systems generally scan all references in a database to see if there is a match to the sample presented. Typically, verification systems are positive claim systems;[10] identification systems can be either positive or negative claim systems; however, due to the nature of those applications (see discussion of positive vs. negative claim systems above) most negative-claim systems (to verify that a subject is not already in the system) are for identification. A middle-ground situation exists where certain members of a user group can choose to be enrolled together (linked). Subsequent system uses by any of the members will cause the system to do comparisons only within that user group, also known as one-to-few matching.

---

[9]A one-to-one relationship in a verification comparison isn't always strictly true. In a verification system one or more candidate samples may be compared against a collection of references that were previously provided by the nominated subject or may be compared to a larger "cohort" set of references from many individuals.

[10]Negative-claim verification systems are theoretically possible, even if rare or nonexistent in practice. An example would be a system to verify that an individual is not a known terrorist.

Closed vs. Open Systems

These categories distinguish systems that are entirely self-contained with regard to biometric data ("closed" systems) from those that must interoperate with other systems ("open").[11] Entirely self-contained systems use enrollment data obtained from the system and do not send any biometric data outside the system. The Disney system for passholders described in Appendix D is such an example. Many systems, however, must receive enrollment data from another system or must pass on collected biometric data. For example, face recognition systems currently being used in immigration control at a number of airports worldwide use enrollment images stored on e-passports. Those images are placed on the passports by passport issuance agencies and are generally based on photographs submitted by the passport applicant. Consequently, the immigration control agency does not directly control the enrollment process. Closed systems do not require adherence to published standards and can adopt processes, software, and hardware solutions optimized for their own application. Open systems must rely on standards usually written for a general class of application and not optimized for any application in particular. Adherence to such standards allows unrelated systems to use each other's data, but it may result in a loss of matching accuracy for all systems involved.

Technology Context

Control of Environment

This refers to control of environmental factors that can affect the system or data subject performance. It may be possible to have an environment that is only partially controlled—that is, where the relevant factors are controlled so that the technology performs as desired. Different technologies are sensitive to different environmental factors. For example, many optical scanning type systems for face, finger, and iris are sensitive to both the visible and infrared spectral areas, so sunlight is an important factor. Silicon-based finger-scanning sensors are not very sensitive to sunlight but are susceptible to static discharge, which makes relative humidity a more important consideration. Other factors affecting the data subject and/or the technology can include temperature, ambient noise level, ambient lighting level and uniformity, geographic location, electromagnetic field noise level, and so on.

---

[11]James L. Wayman, Technical testing and evaluation of biometric identification devices. In Jain et al., eds. *Biometrics: Personal Identification in Networked Society*, Boston, Mass.: Kluwer Academic Publishers (1998).

Passive vs. Active Technologies

Passive technologies do not require the data subject to undergo a controlled interaction with the system although they may require cooperation from a data subject whether the subject is knowingly or unknowingly providing that cooperation, such as looking in a particular direction. Most facial and iris recognition technologies are considered passive but require a certain amount of cooperation by users, since at present they must look in a particular direction to be scanned successfully. On the other hand, gait recognition might be considered completely passive. Voice recognition applications could also be considered passive if they are based on regular conversational speech without subject interaction. Active technologies are those that require direct human interaction such as speaking a particular phrase or positioning a finger, hand, or head in the correct location. Some technologies might be considered semipassive, active in that a subject must position a feature in a certain orientation but passive in that there is no need to actually contact any surface and orientation need not be at all precise.

Covert vs. Overt Technologies

Covert technologies are those that are designed to be used without the user's knowledge. Overt technologies actively involve the user.

Performance Context

Throughput Requirements

Throughput refers to how quickly data subjects can be successfully processed by the system. This includes both successful and unsuccessful attempts at recognition. How long will a cooperative subject be allowed to present a feature before a timeout is called? How many retries will a user be allowed? Most applications have some degree of throughput required before the system becomes economically or procedurally prohibitive, and all applications should include the time required to process both correct and false matches that lead to application exception processing, since these cases tend to take more time and manual intervention than application acceptances.

Error Rate Requirements

Error rates, technically more properly referred to as error proportions, measure the frequencies of errors relative to the number of adjudications that could produce errors of the corresponding type. This term receives by far the most public attention and is likely also the most misused term. There are multiple components that contribute to system inaccuracies, and only a few are technology dependent. These are typically contributions from signal noise and background noise, but typically the largest components are the human interaction and environmental components. Error rates have direct impact on throughput, since false rejections take significantly longer to process than acceptances (whether true or false). Applications that require high throughput should aim for low false rejections; however, care must be taken in the application's security analysis to balance low false rejections with the potential for an increase in false acceptances.

INTEROPERABILITY

Interoperability is a factor to consider when designing almost any kind of system. It must of course be considered when discussing data exchange between systems but must also be considered for evolving systems that retain and reuse data collected over time. In other words, even closed systems might have to interoperate with multiple generations and instantiations of themselves as they change over time. Interoperability also plays a role at the subsystem level, when systems are composed of vendor software or hardware. In the biometric system context, such components may include fingerprint matcher components, segmentation software, and minutiae detectors.

In general, standards help to promote interoperability. However, there are times when the use of a standard format in preference to a proprietary format can be detrimental and potentially limit functionality or flexibility.[12] For biometric systems, sensor interoperability, discussed below, poses some specific challenges.

Sensor Interoperability

Sensor interoperability refers to the compatibility between an enrolled biometric reference and a test sample, acquired using different sensors. In some systems, it is assumed that the two samples to be compared were

---

[12]For example, the NIST Minutiae Interoperability Exchange Test 2004 compared the performance of proprietary systems against the performance of template data in standard format generated and matched by different vendors. In these tests the use of standards-based formats for single-finger matches was inferior to proprietary systems. Changing the system to use two-finger matching adequately compensated for the reduced accuracy but remained inferior to proprietary two-finger operation. Of course, the operational issues encountered when switching from collection of one fingerprint to two fingerprints may militate against the use of the latter two-fingerprint method to improve accuracy.

acquired using the same sensor—or at least the same type and vintage of sensor. However, improvements in sensor technology and reduction in sensor costs means that enrollment and test samples are often obtained using different sensor types. This may also happen if a sensor manufacturer goes out of business and support is no longer available for a line of sensors. In a large, distributed system such as the FBI's Integrated Automated Fingerprint Identification System (IAFIS), remote sites use a variety of end-point systems and may collect samples using a variety of certified sensors. In short, it is possible in almost any biometric system that test samples will not be collected using the same sensor as used in enrollment. (In criminal justice systems, there is no known change in error rates when each booking site selects whichever certified scanning system it wants.) It has been observed that the matching performance drops when the reference and test samples for fingerprint, iris, and voice are acquired using different sensors rather than the same.[13]

There are several reasons for this degradation in matching performance: (1) change in the sensor resolution and its operating behavior; (2) change in the sensor technology; (3) change in the user interface, and (4) changes in operational environment that have an impact on sensor performance. In the first case, while the underlying sensing technology remains the same—in, for example, an optical total internal reflection (TIR) fingerprint sensor—the image resolution (say from 300 by 300 to 500 by 500 pixels per inch) and/or the signal to noise ratio (SNR) of the sensor may change. In the second case, the two sensors providing reference and test samples may be based on completely different technologies (for example, one may be an optical TIR fingerprint sensor and one a capacitive solid state sensor). Of the factors listed, the second is more problematic because the change in sensor technology may decrease compatibility between enrolled references and test samples. There are multiple examples of different sensing technologies for a given modality. For example, sensors for fingerprints can be based on optical, capacitive, thermal, pressure, ultrasound, or multispectral technologies. Some touchless three-dimensional fingerprint sensors are being developed as well. Differences in how a user must interact with a sensor may introduce variations in sample coverage area and distortion. Similarly, two-dimensional images for face recognition can be captured in visible color, infrared, and thermal, as well as range (depth). Three-dimensional face images that

---

[13]See, for example, A. Ross and R. Nadgir, A calibration model for fingerprint sensor interoperability, P. Flynn and S. Pankanti, eds., *Proceedings of SPIE Volume 6202, Biometric Technology for Human Identification III* (2006); Modi et al., Statistical analysis of fingerprint sensor interoperability performance, *Biometrics: Theory, Application, and Systems* (2009); and International Biometric Group, Independent testing of iris recognition technology, *Report to Department of Homeland Security* (May 2005).

capture the depth image are also being used for face recognition; these so-called range sensors capture a different face modality than the usual two-dimensional intensity or texture images captured by charge-coupled device (CCD) cameras.

Sensor interoperability is a major concern in large biometric installations, since it is expensive and time consuming to re-enroll a large number of subjects as the technology evolves and access to the subject population may become limited. In some cases, re-enrollment may be unavoidable and should be viewed as part of upgrading the infrastructure. To the extent possible, of course, representations of traits that perform well across existing and anticipated technologies are desirable.

Human Interface Interoperability

One aspect of interoperability is the development of standardized human interfaces that would allow the data subject to know what to expect when interacting with a biometric system and how to control the recognition process. Although other technology interfaces such as are found in automatic teller machines, automobiles, televisions, and self-service gasoline pumps have a level of standardization that allows transferring experience gained with one system to other systems, little has been done in this area for biometrics, and these mass-market interfaces can confuse even experienced users on occasion. More standardized user interfaces coupled with broader human factors testing would contribute to greater maturity in all biometric applications.[14]

SYSTEM LIFE-CYCLE ISSUES

Biometric systems that are large in scale and that are expected to persist and be used for more than a short period of time face the same challenges as other large-scale technology implementations.[15] Software and

---

[14]For more on usability and biometric systems, see "Usability and Biometrics: Ensuring Successful Biometric Systems," available at http://zing.ncsl.nist.gov/biousa/docs/Usability_and_Biometrics_final2.pdf. In particular, the report notes:

> In order to improve the usability of biometric systems, it is critical to take a holistic approach that considers the needs of users as well as the entire experience users will have with a system, including the hardware, software and instructional design of a system. Adopting a user-centric view of the biometric process is not only beneficial to the end users, but a user-centric view can also help to improve the performance and effectiveness of a system.

[15]In the case of biometric systems, scale can refer to the number of sensors in the system, the number of comparisons being performed for a given unit time, the number of users (administrators and data subjects), the geographic distribution of the system, the potential number of data subjects, or any combination of these factors. The point here is that there is

hardware have limited lifetimes and become obsolete. The maintenance and technology refresh requirements of biometric systems can be poorly understood and neglected as a result. Most biometric systems have basic needs, such as cleaning of the lenses or plates, sensor recalibration when the components or the environment changes, software updates to keep pace with hardware changes, and so on. Evolution of large-scale systems while in use requires careful pretesting to verify the ability to migrate from the old technology to the new and have both coexist in the system simultaneously. Some system components may change without maintaining backward compatibility. Technologies will change significantly over the expected lifespan of a system, and biometric components may need to be updated. In general, biometric systems may be similar to other computer-based systems in that useful lifetimes cannot be expected to surpass 5 to 7 years without becoming obsolete.

Unlike some computer-based systems, however, biometric systems are likely to have a critical hardware component, making upgrades and replacements more logistically challenging than, say, pushing software updates to a networked information system. For example, consider the simple case of a single fingerprint sensor, deployed to provide data security (rather than convenience) by controlling access to a laptop computer. In normal operation, this application will be limited to repeated private interactions between the owner and machine for the machine's full life. However, a sensor can fail from simply wearing out, from physical damage due to rough usage, from a dirty or otherwise inimical environment, or from intentional damage by an unauthorized user who obtains control of the machine. The fingerprint acquisition and/or fingerprint matching software, or the file with the enrolled biometric template—a software issue—can also be corrupted. What happens if the system can no longer be used to recognize the user? Is the user simply cut off, or are there alternative access options? And how are those alternative access options made clear to the user later in the system's life cycle in a way that does not compromise the security that the system was deployed for in the first place?

Quality control, especially in large-scale systems, is critically important. When engineering a biometric system, planning how to ensure continued high-quality performance is key. At the same time, mechanisms are needed to detect and accommodate degraded performance, should it occur. Related to this are issues of scalability, as mentioned previously.

---

a significant difference between a very localized, small-scale application (such as facilitating quick entry for members of a local gym) and much larger systems (such as border control or social welfare systems). This section focuses primarily on the technology and engineering challenges associated with large-scale biometric systems.

Can performance be maintained as the system scales up either in terms of the size of the user population, number of sensors, geographic distribution, size of the search space, or some other dimension? For a given system, what characteristics must be monitored and compensated for as a system grows? Possible areas of improvement include matcher improvement, collecting more data—for example, 10 fingers instead of 2 or other types of multibiometrics—and so on.

Security is also a life-cycle consideration. As described in Chapter 1, security considerations are integral to the design of a biometrics system. Risks are not static, and changes in attack methods, potential exposure or compromise of biometric data, and the emergence of new vulnerabilities to threats need to be assessed periodically. Consider the example of access to a laptop controlled by a fingerprint sensor cited above. In high-confidence systems, sensor replacement, for example, poses a risk because a compromised sensor used to replace a legitimate sensor could allow an attacker to access the system. A draconian solution to this security problem would be to automatically monitor sensor performance, letting sensor failure or replacement initiate a data overwrite or physical destruction of the storage medium or of cryptographic keys used to encrypt storage on the machine. However, this approach could be detrimental to the owner unless exceptionally stringent backup systems were in place. Alternatively there could be a backup computer activation mechanism and password authentication for the owner or a computer locksmith. This would partially spoil the purpose of the fingerprint device, however, and the entrusted locksmith would have to be trustworthy. Furthermore, as with computers that are not biometrically protected, stored data lacking sufficient encryption might be accessed by physical removal of the data storage medium. The idea is not to examine or critique various approaches for dealing with such contingencies but to point out that even in this simple situation, numerous issues that would arise during the life cycle need to be resolved during the system design and engineering phase. Future-proofing, to the extent possible, and paths to technology refreshment can encompass issues such as software and hardware modularity—including sensors and matchers—as well as common interfaces and interoperable data formats.

TEST AND EVALUATION

Biometric systems are almost always components of larger systems designed to perform a business process. Testing and evaluating these larger systems depends on the biometric system, application, concept of operations, functional and performance requirements and on many deployment-specific factors that must be considered, including conformance to standards. (See Appendix E for a brief overview of the biometrics standards landscape.) The factors include the environment, maintenance, subject population, subject education, subject cooperation, and policy.

Testing and evaluating of biometric systems can serve multiple purposes, including estimating performance under real-world usage of a biometric system; developing or examining metrics related to biometric systems; determining conditions that affect performance; supporting system procurement; accepting a system at the time of delivery to an operational site; and letting developers and program managers know about opportunities for system improvement. Testing can occur at many levels and might involve presenting original samples to a biometric system and recording the results, or might involve only presenting a standardized database of biometric characteristics to feature extraction and matching algorithms. More generally, testing and evaluation are broad terms. Aspects of systems that could be measured include technical performance and accuracy, throughput, interoperability, conformance (to standards or requirements), reliability, availability, maintainability, security and robustness against vulnerabilities, safety, usability, and public perceptions and acceptance. As this chapter describes, work in all of these areas is needed, although progress on the first three is further along than the others. The testing and evaluation of particular aspects of systems can be aimed at either assessing the performance of a particular system in isolation or comparing the performance of similar systems. For examples of test and evaluation practice in three real-world examples (the FBI's IAFIS, Disney's entrance control system, and the U.S. Army's BAT system), see Appendix D.

NIST divides testing of biometric systems into four general categories: conformance, scenario, interoperability, and technology testing. In addition, a full testing and analysis regimen includes operational and usability testing.[16] Conformance testing addresses technical interface and data interchange standards. Scenario evaluations use volunteers to model data subject interaction with a system in a laboratory environment designed to model a target application. These scenario tests include system level testing focused on performance as well as human centered biometric sample collection testing. Interoperability testing addresses sensor interoperability and modality-specific templates and matchers. Technology testing evaluates feature/model extraction and comparison algorithms using a

---

[16]See the NIST National Voluntary Laboratory Accreditation Program (NVLAP) Handbook 150-25 on Biometric Testing, which is available at http://ts.nist.gov/Standards/Accreditation/upload/NIST-HB150-25-2009.pdf and http://ts.nist.gov/Standards/Accreditation/bio-lap.cfm. As of July 2009, laboratory accreditation under NVLAP is available only for conformance testing and aspects of system testing under scenario testing.

standardized database of biometric samples; sensors are tested using standardized test materials. Operational testing looks at an existing system in situ. Usability testing evaluates the effectiveness of interaction between the users and operators of a biometric system. Evaluation of biometric systems includes all of the factors noted above as well as the analysis and assessment of technology, scenario, and operational test results and of the training process. An evaluation plan generally begins with the purpose, application, concept of operations, functional and performance requirements, and development of relevant metrics.

The data used in test and evaluation of biometric systems is an important piece of the T&E process. Biometric subsystem testing can be conducted using found data that mirrors some aspects of the application or conducted on an operationally deployed system (including recording the operational data for later evaluation).[17] Depending on the purpose, there are best practices, such as the ISO/IEC 19795 series of standards for conducting tests and evaluations of biometric systems.[18]

Testing and evaluation are important throughout a system's life cycle. While it may seem obvious to test typical deployment use of a biometric system, other phases of system use merit evaluation as well. Some biometric systems require a training phase—one where biometric samples are presented to the system and models are built to be used during recognition. This phase might be evaluated with a properly designed test. Another type of evaluation is a core technology evaluation to compare various matching algorithms on a common task using common data. Yet another example is to predict real-world performance by evaluating performance in a laboratory scenario.

Regardless of which aspects of the system or system life cycle are under scrutiny, the testing discussed in this report is considerably more sophisticated than brute force tests aimed at quickly stressing a system or device to the point of failure. The evaluation approaches discussed here are intended to scientifically evaluate the performance of presumably properly working devices in various dimensions. At the same time, as

---

[17]"Found data" is a term that applies to data collected for some other purpose by someone who is willing to make the data available for testing. Use of "found" biometric data is generally problematic because of Institutional Review Board concerns over limiting data use to the application for which it was collected. Nonetheless, there are several "found" databases available for testing.

[18]Work along these lines includes efforts undertaken by the British government since 1999, NIST Speaker Recognition Evaluation since 1994, and ISO/IEC JTC1 SC37 since 2002. There is also NIST's emerging National Voluntary Laboratory Accreditation Program (NVLAP) for biometrics (see the NIST NVLAP Draft Handbook 150-25), which establishes a mechanism for biometric testing laboratories that can be certified for different types of product tests; however, the methods for different products and applications had not yet been specified at the time of this writing.

will be seen in the brief descriptions of the biometric system tests below, abstracting to general principles for evaluating a biometric system and developing appropriate operational tests and evaluative techniques based on those principles can be a challenge. Even with significant testing, great caution is warranted in generalizing from one system to another or extrapolating behavior of a system across environments or user populations.

Usability Evaluations

Biometric system evaluations have historically been centered on error rate and throughput estimations, and they have tended to neglect usability considerations and acceptance testing. However, for all systems, even covert, both measured error rates and throughput are dependent on human interaction with the system. For example, the NIST testing taxonomy mentioned previously emphasizes algorithm testing in technical tests and does not significantly focus on human interaction. NIST's recently initiated assessment of human interactions with the biometric system offers a real opportunity to enhance biometric system testing.[19] The NIST efforts focus on the user as data subject, that is, on the individual who is presenting a sample to the system for recognition. This could be broadened to include system operators and system administrators, and, in some cases, systems owners as users.[20]

Test and Evaluation Standards

Biometric testing standards have evolved to address various forms of testing. Biometric performance testing and reporting of international standards, published as the ISO/IEC 19795 series of standards, evaluate biometric systems in terms of error rates and throughput rates.[21] Metrics for the various error rates in biometric enrollment, verification, and identification are specified. Recommendations are given for evaluating performance through planning the evaluation; collection of enrollment,

---

[19]Results of the NIST usability tests are publicly available on the biometrics and usability website at http://zing.ncsl.nist.gov/biousa/.

[20]A recent workshop hosted by the NRC Computer Science and Telecommunications Board examined the broader issue of usability, security, and privacy. See National Research Council, *Toward Better Usability, Security, and Privacy of Information Technology: Report of a Workshop*, Washington, D.C.: The National Academies Press (2010).

[21]The introduction to the standard notes that "technical performance testing seeks to determine error and throughput rates, with the goal of understanding and predicting the real-world error and throughput performance of biometric systems. The error rates include both false positive and false negative decisions, as well as failure-to-enroll and failure-to-acquire rates across the test population." It should be noted that these standards do not fit the AFIS/ABIS system benchmarking and acceptance testing that governments perform.

verification, or identification transaction data; analysis of error rates; and reporting and presentation of results. The principles apply to a wide range of biometric modalities, applications, and test purposes and to both offline and online testing methodologies. These principles are aimed at (1) avoiding bias due to inappropriate data collection or analytic procedures; (2) providing better estimates of field performance for the expended effort; and (3) clarifying the extent to which test results are applicable.[22]

Performance Assessment and Evaluation

Performance—which goes beyond whether the system returns adequate recognition results for a given application—can be a critical feature of biometric systems, particularly at scale. In this sense, performance encompasses not just error rates but also throughput, reliability, and other features crucial to system success. For example, systems that are engineered so that performance can be dynamically monitored during testing and deployment should offer system administrators performance data throughout the operational life cycle. Similarly, the system development and testing communities would have an opportunity to work on ways to tune the operation of a biometric system to maintain performance metrics at acceptable levels in the face of a changing load and environment. The performance of systems depends on the performance and interrelationships of their components (including the data subjects and human operators). Component performance is best measured and understood by studies that limit the number of factors involved through experimentation or tightly constrained observation that adequately model the target application.

Modeling, including simulation, also may be useful, but it must be grounded in observation. Even when the performance of a system depends in a simple manner on the performance of its components operating independently, only holistic study of the system can confirm that dependence. An understanding of performance and realistic planning for improvement require both analytical studies of individual system components and holistic studies of full system operation. Consequently, performance statistics cannot be accurately predicted for systems orders of magnitude larger than those previously studied or that have very different hardware/software/user interfaces.

Characteristics that limit system performance or opportunities to improve it may reside with individual components or with the manner in which they interact within the system. However, improving the perfor-

---

[22]Several organizations and institutions perform biometric testing, and there are several periodic large-scale tests in which many biometric vendors participate.

mance of one component sometimes degrades the performance of a larger system in which it resides.[23] There is no way to definitively determine the impact of component changes on system-level performance until the components have been inserted and the system is tested. System performance may depend critically on factors such as the current performance levels of other components, whether the performance characteristics of some components depend upon the outputs of other components, the nature and quality of system input material, and the characteristics of the environment in which the system operates.

---

[23]Upgrading a face image detector in a facial recognition system is one example of how improving performance of a component can degrade the system. If the system can find more faces in more poses, then the number of off-angle and partial faces sent into the comparison process increases. This suboptimal data situation can lead to poor performance. Moving from lossy to loss-less compression is another example where a local improvement can degrade overall system effectiveness. See, for example, NIST, Effect of image size and compression on one-to-one fingerprint matching, NISTIR 7201 (February 2005), and J. Daugman and C. Downing, Effect of severe image compression on iris recognition performance, *IEEE Transactions on Information Forensics and Security* 3:1 (2008).

3

Lessons from Other Large-Scale Systems

The many large-scale biometric systems in use today are deployed in a broad range of systems and social contexts. The successes and failures of these biometric systems offer insights into what can be learned from careful consideration of the larger system context, as well as purely technological or component-level aspects, during planning. Common characteristics of successful deployments include good project management and definition of goals, alignment of biometric capabilities with the underlying need and operational environment, and a thorough threat and risk analysis of the system under consideration. Common contributors to failures include the following:

- Inappropriate technology choices,
- Lack of sensitivity to user perceptions and requirements,
- Presumption of a problem that does not exist,
- Inadequate surrounding support processes and infrastructure,
- Inappropriate application of biometrics where other technologies would better solve the problem,
- Lack of a viable business case, and
- Poor understanding of population issues, such as variability among those to be authenticated or identified.

Many of these factors apply in any technology deployment, biometrics-related or not.

While much can be learned from studying biometrics systems, it

76

seems appropriate, given their scale and scope, to consider whether the biometrics community can learn lessons from large-scale systems that have been deployed in other domains. This chapter explores some of the technical/engineering and societal lessons learned from large-scale systems in manufacturing and medical screening and diagnosis. In each case, the discussion points out useful analogies to biometric systems and applications.

MANUFACTURING SYSTEMS

Manufacturing systems convert initial materials into finished products that must meet quality specifications. Each step in the conversion may consist of a complex process sensitive to multiple characteristics of the input materials and processing conditions. Each step also represents an economic investment; modifications to the process that can achieve equal or higher quality at lower cost are every company's goal. Production-line systems have been studied systematically since before World War II from the perspectives of industrial engineering, statistics, experimental design, operations research, and quality control. Insights gained from the study of such systems have been generalized to better understand and improve the performance of systems for product development and other industrial processes and to facilitate improvements in corporate management.

A simple example, used in a 2005 briefing to the study committee by Lynne Hare of Kraft, Incorporated, is the development of a new sensor for a manufacturing production line. The process begins with identifying the business need for the sensor and proceeds through its implementation and then deployment in the production line. The stages include explicit translation of the business need into the scientific requirements for the sensor, fabrication of a prototype sensor, preliminary (static) testing, formal static and dynamic testing, pilot installation and testing, and production line implementation and validation. The process never ends, because revalidation is scheduled at periodic intervals. At each stage of testing and data collection, the information obtained may send the development process back to an earlier stage to correct any observed deficiencies and improve robustness of the sensor to varying conditions.

This example can be interpreted directly or as analogy. Directly, it gives a model for developing and implementing devices required by any biometric system to sense biometric traits, for example, fingerprint scanners, iris scanners, and audio recorders. There is also an analogy between development and validation of a sensor and the development and implementation of a biometric system. In this analogy, the multiple levels of testing—preliminary static, formal static and dynamic, and production line testing—are counterparts to technology, scenario, and operational evaluations of a biometric modality. Motivations for these three levels of testing in the sensor development environment can be informative for the development and testing of an entire biometric system.

Additionally, a biometric system may be considered as a production line, the inputs as individuals presenting for recognition, and the output as a series of decisions that will achieve a high quality, reflected in low values of the false match rate and false nonmatch rate, and in a ratio appropriate for the system's intended purpose(s). When a biometric system is looked at in this way, it can be seen that the methods of industrial engineering and statistical quality control can be applied to achieve system quality.

At least three fundamental insights into managing industrial processes are also relevant to biometrics. The discussion below paraphrases selected core concepts from the work of Deming, Shewhart, Box, and their many successors.[1]

One insight is that careful articulation of requirements, preferably in measurable terms and derived from an end product or process, is exceptionally important to the successful development and implementation of component parts. In the case of a production line, for example, a requirement might be for the sensor to respond reliably and repeatedly only to stimuli in the desired range and measure stimuli accurately, under conditions in the production environment. The range of stimuli, sensor sensitivity and resolution, and resistance to environmental disturbances must be accurately specified during the design process in order for the sensor to properly identify defective units, which is its ultimate purpose. By analogy, biometric system design should be driven by clear objectives for the recognition task in the context of the broader application rather than merely by the existence of an attractive technology.

A second insight is that a scientific approach is invaluable to understanding systems, particularly the interrelatedness of system components. The hallmark of the scientific approach is exploration through both theory and data. The performance of complex systems can be often improved by identifying and correcting bottlenecks or other localized problems whose negative effects may not have been fully perceived and articulated. Such problems and other aspects of interrelatedness and individual component performance can be identified by planned data collection guided by careful theorizing about the system. Such data may be collected by observation, as exemplified by the use of statistical control charts in a

---

[1]George E.P. Box and Owen L. Davies, *The Design and Analysis of Industrial Experiments*, Edinburgh: Oliver and Boyd (1954); Walter A. Shewhart, *Economic Control of Quality of Manufactured Product*, Milwaukee, Wisc.: Quality Press (1980); W. Edwards Deming, *Out of the Crisis*, Cambridge, Mass.: MIT Press (1986).

production line, or by direct experimentation on the system itself. In such experimentation, system inputs or conditions are systematically modified to learn their effects on the functioning of the system and the quality of its output. Such experiments can be carried out from time to time or can be built into the system itself. Evolutionary operation (EVOP), an example of the latter, refers to the regular alteration of baseline system parameters by small amounts during production runs. The changes made are too small to disrupt system operation, and the system is run with these changes in place just long enough to assess the effects on product quality and other aspects of performance. Changes that most improve performance may then be retained, and the process continued from new baseline values of parameters. Iterations of such experimentation gradually nudge the system toward optimal parameter values by exploring nonlinear regions of the "response surface" that relates performance to different combinations of parameter values.

A third insight, stressed in statistical quality control and one of the four pillars of Deming's "system of profound knowledge,"[2] is the importance of understanding background variation in system performance and identifying separable contributors to it. The first and foremost meaning of "understanding" in this context is recognition that systems exhibit natural variability due to random influences, and that inordinate reaction to such short-run variability is often wasteful and of little benefit. Although dramatic but relatively brief slumps and streaks are a major source of discussion by sports analysts and some stock traders, basing major decisions on such brief events rarely leads to prosperity for a baseball team or an investor. A deeper level of understanding develops from the awareness that random variation in output typically comes from multiple sources that persist even as its momentary influences fluctuate. In technical parlance, these sources and the measures of their strength are often referred to as components of variation (or variance), and in industry parlance as "the voice of the process." In a manufacturing process they might include variability in raw material batches, calibration drift of instruments guiding system processes, problems with machinery maintenance, and human error. Reducing the variation from such common sources can improve product quality over the long run. Some variation arises, however, from "special" sources that would not be expected to recur and against which changes to the system can do little to protect. Thus, identification and reduction of the largest components of variance from common sources is generally accepted as critical to quality improvement in industrial production systems. Some version of Deming-Shewhart plan-do-study-act

---

[2]See W.E. Deming, *The New Economics for Industry, Government, Education*, Cambridge, Mass.: MIT Press (2000).

(PDSA) cycles is generally used to bring a scientific approach to bear on this task.

The insights sketched above apply to biometric recognition systems no less than to any other systems. But since they provide an approach rather than a prescription for learning about and improving systems, their implications will vary greatly according to context. Moreover, for many operational biometric systems the "ground truth"—that is, the "correct" answer in terms of system objectives—is indeterminate for many transactions. The approach described above is invaluable in developing such systems. The emphasis on examination of process variables in an operational mode is potentially very helpful. Its potential benefits are even greater if challenge experiments can be superimposed on the operational system. There is substantial precedent for such challenge experiments in other contexts, including evaluations of Internal Revenue Service tax assistance and Transportation Security Administration airport passenger and baggage screening.

So, independent of the particular biometric modality and its application, the following lessons can be drawn from the experience and methodologies that have evolved in industrial production:

• System objectives must be clarified at the outset if the system is to be designed efficiently and if the ability to evaluate system performance is to be preserved. In particular, the often-interrelated but distinct goals of improving convenience, controlling access, detecting threats, lowering costs, tracking and managing employees, and deterrence must be distinguished and prioritized in system planning.

• The operational environment, including the range within which environmental characteristics and characteristics of the populations presenting to the system will vary, should be anticipated as much as possible in systems development. This includes consideration of operation under routine conditions; under unusual conditions unconnected to any specific threat; under realistic threat scenarios for attempts to defeat the system at the individual level; and under realistic threat scenarios for penetrating, degrading performance, or shutting down operations at the system level.

• To the extent that systems are mission-critical, large-scale, and addressed to national security, controlled observation at the operational level, including ongoing challenge experimentation, is essential. In routine operation, many errors are likely to occur in which an individual making a true recognition claim is at first erroneously restricted but the mistake is later discovered and corrected, at which point these errors become visible and available for analysis. However, when an individual gains unauthorized access because, for example, a false claim of identity goes undetected, the error may remain undiscovered for a long time.

Challenge experiments, which observe and compile system responses to inputs representing (1) typical experience, (2) variations in conditions, (3) difficult presentations requiring adjudication or systems adaptation, and (4) attack modes, are the best way to identify the potential for such errors and ways to prevent them. Erroneous rejections of true recognition claims and erroneous acceptances of false claims should be documented and subject to rigorous fault analysis, just as would take place in the case of an investigation into a transportation crash. Such analysis should include comparison with a sample of correct recognitions used as controls in order to distinguish factors predisposing to errors from those predisposing to correct decisions.

• Studies of system behavior, including those attempting to discover and reduce the largest contributors to system error and the most variable components of intermediate products that contribute to recognition decisions, may be as revealing and helpful for biometric systems as they have been for systems involving other repeatable processes.

MEDICAL SCREENING SYSTEMS

Medical screening systems collect diagnostic information, generally in a staged sequence, in an attempt to locate individuals with an undetected disease that can be more effectively treated early in its course. The input to such a system is data from a population of individuals, some with disease but most without. Results of the stages generally are classified as positive or negative, and only individuals who test positive at each stage are labeled by the system as having the disease. Consider the following simplified (and not necessarily medically realistic) view of a prostate cancer screening system. Screening is initiated by a digital rectal examination. The patient with a normal exam is not screened further. Abnormal palpation results, however, are followed by a prostate-specific antigen (PSA) test. Patients with a PSA level below a certain point are not screened further. When the PSA level is at or above this point, the prostate is biopsied. When the pathologist finds the biopsy to be negative for cancer, the patient is so informed. When the pathologist finds it to be positive for cancer, the diagnosis is considered to be established, and the patient is referred for consideration of treatment alternatives. The alternatives, and indeed the importance of any treatment, may vary depending on age of the patient, stage of the cancer, and rate of progression, which may often be determined by a watch-and-wait period.[3]

Each component test of this progression will detect some prostate

---

[3]The point here is that the disposition of a medical screening result may vary as a function of patient factors, and what happens after that is, nonetheless, appropriately viewed as a system output. Similarly, the disposition of a biometric recognition (or lack of recognition)

cancers and miss others, its false negatives. Some men without prostate cancer, perhaps with another disease such as benign prostatic hyperplasia (BPH), will be classified positive at one or more steps. The proportion of cancers detected is known technically as "sensitivity" and the proportion of prostate cancer-free individuals classified as negative is known as the "specificity." The complementary proportions—that is, the proportion of prostate cancers missed and the proportion of men without cancer who are identified as positives—are called the false negative and false positive rates. These are analogous to the false match and false nonmatch rates in a biometric recognition application. Note that each component of the screening system will have its own values of these numerical characteristics describing the performance of that component and that another set of values characterizes the performance of the screening system overall. In practice, the true values are generally unknown, but hypothesized or estimated values coupled with well-established mathematical relationships can provide useful guidance for screening policies.

Screening systems have been extensively studied in a medical context. Their general characteristics are well understood, but their specific performance levels may be unclear. The following lessons are among those that have been learned:

• Individual components in general usage are rarely as sensitive and specific as the components when they were under development because tests are usually developed and evaluated by researchers exceptionally skilled in their use on subjects whose states of health or disease are well known.

• The value of each component to the screening system is determined not just by its individual properties but by the information it contributes in addition to the contribution of the other components. For instance, confirming the result of a test by repetition is less valuable than confirming it by a different test that screens for a different disease marker.

• Limitations of individual components can vitiate the effectiveness of other components. For instance, in the system described above, a pathologist who cannot detect true prostate cancer renders the accuracy of earlier components in the sequence virtually irrelevant.

• Effectiveness of a system is highly population-specific, even when the system's overall sensitivity and specificity are exceptionally high. This is easily seen by considering a screening system implemented in a population from which the disease in question is absent. No matter how high the

---

might vary according to situational and subject factors; because the consequences of the system's results affect system output, they are important in evaluating a biometric system.

sensitivities and specificities of the system components and of the system as a whole, all positives will be false positives and the screening system will provide no health benefit.

• In view of the preceding item, the performance of a system is best represented by its population-specific predictive values—that is, the proportion of screen-positive individuals who truly have the disease (positive predictive value) and screen-negative individuals who truly do not have the disease (negative predictive value). Alternatively, the ratios of screen-positives with the disease to screen-positives without the disease and the ratios of screen-negatives without the disease to screen-negatives with it, may also be used to represent performance. These measures combine information on the accuracy of the testing (sensitivity and specificity) with information on the composition of the population, since both are critical to determining whether screening is informative.

• The ability of a system to detect disease and the importance of detection may vary by characteristics of the disease and the patients in whom the disease occurs. For instance, screening is more likely to detect slowly progressing (indolent) than rapidly progressing (aggressive) disease, because the symptom-free period is longer for the former. But sensitivity is less important in detecting indolent disease, because subsequent rounds of screening may detect it before it has progressed much further. In the case of prostate cancer, elderly men with the indolent form may be more likely to die from something else before the cancer kills them.

These observations are general, and the analogy to biometric systems is imperfect. They do, however, have some implications for biometric systems:

• Laboratory and scenario testing are apt to underestimate field error rates of biometric applications.

• Combinations of independent or minimally dependent characteristics and processes generally incorporate more information, and thus offer higher potential for improved performance, than combinations of more correlated components. Hence, in biometrics systems design, independent features, components of multimodal biometrics, and components of decision-making scores are preferable to combinations of correlated alternatives of comparable cost.

• A poor adjudication process, or an ineffective backup process for dealing with failures-to-acquire (see Chapter 2) in a biometric system, may negate the benefits of good error rates in the basic biometric technology.

• Biometric technologies must be calibrated to the environment and population in which they will be implemented. For instance, one might expect different operational characteristics for biometric border-control systems using identical technology on the Mexican border with Texas and the Canadian border with New York, in part because the frequency of attempted illegal border crossings in these places is so different.

• System performance characteristics may vary by major population subgroups and by the types of challenges presented to the system. Extrapolation of technological or system performance characteristics across settings or challenges—for example, from (1) laptop access control to auto theft control to border control or (2) from illegal immigrants to narcotics smugglers to terrorists—is unlikely to be reliable.

4

Cultural, Social, and Legal Considerations

Biometric systems assume and require an intimate relationship between people and technologies that collect and record the biological and behavioral characteristics of their bodies. It is therefore incumbent upon those who conceive, design, and deploy biometric systems to consider the cultural, social and legal contexts of these systems. Not attending to these considerations and failing to consider social impacts diminishes their efficacy and can bring serious unintended consequences.

The key social issue surrounding biometrics is the seemingly irrevocable link between biometric traits and a persistent information record about a person. Unlike most other forms of recognition, biometric techniques are firmly tied to our physical bodies. The tight link between personal records and biometrics can have both positive and negative consequences for individuals and for society at large. Convenience, improved security, and fraud reduction are some of the benefits often associated with the use of biometrics. Those benefits may flow to particular individuals, corporations, and societies but are sometimes realized only at the expense of others. Who benefits at whose expense and the relative balance between benefits and costs can influence the success of biometric deployments.

The efficacy of a biometric system can be affected by the cultural, social, and legal considerations that shape the way in which people engage and interact with these systems. People's deliberate choices about whether and how to engage and their inadvertent actions both affect system performance. For example, some people may choose not to place their fingers on a fingerprint scanner for fear of contracting a disease or may be unable to do so because long fingernails are highly valued by their social group. Similarly, some people may avoid having their photographs taken for a face recognition system because of concerns over how the images will be used; others will avoid this owing to concerns about the absence of customary adornments to the face (for example, scarves). In both cases system performance may be compromised.

The proportionality of a biometric system—that is, its suitability, necessity, and appropriateness—in a given context will have a significant effect on the acceptability of that system.[1] The societal impact of such systems will vary significantly depending on their type and purpose. For example, the use of iris scanning to control access to a local gym and of finger imaging to recognize suspected terrorists at international borders are likely to differ, both for the individuals being scanned and the broader community. The potential impacts on particular social groups and thus their receptions by these groups may also vary dramatically due to differences in how the group interprets the cultural beliefs, values, and specific behaviors. Imposing facial recognition requirements to enter a store or workplace may limit the shopping and work options available to individuals who consider photographs of faces inappropriate, creating barriers to social activities.

This chapter explores such considerations in four areas: biometric systems and individual participation, potential impacts on society of biometric systems, legal considerations with respect to biometrics, and data collection and use policies.

INTERACTION BETWEEN BIOMETRIC SYSTEMS AND INDIVIDUALS

System performance may be degraded if social factors are not adequately taken into consideration. These factors are of two types, those that motivate and those that facilitate participant engagement with the system.

Motivating Participation by Individuals

As a rule, peoples' willingness to participate in a system and their commitment to it depend on their understanding of its benefits. For example, a biometric system that allows convenient access to a worksite might be perceived as beneficial to individuals by relieving them of the necessity

---

[1]European Commission, Article 29. The Data Protection Working Party observes that proportionality has been a significant criterion in decisions taken by European Data Protection Authorities on the processing of biometric data. Available at http://ec.europa.eu/justice _home/fsj/privacy/docs/wpdocs/2003/wp80_en.pdf.

to carry an ID card. On the other hand, a biometric system that tracks day-time movement of employees might be perceived as primarily beneficial to the employer and as undermining the employee's personal freedom. In some instances, ancillary inducement, such as monetary reward, may be required to obtain a desired level of participation.

Participation may also be motivated by the possibility of negative consequences for nonparticipation—for instance restrictions on access to locations or services (perhaps entry to the United States), requirements to use a much more lengthy process for a routine activity (for example, to open a bank account), and even the threat of legal action (for example, the requirement to enroll in a biometric system in order to maintain legal alien status). Nonparticipation may also subject individuals to social pressure and/or prevent them from joining some collective activities.

Willingness to participate also may be influenced by concern that system uses will change over time (often referred to as "mission creep"), perhaps becoming less benign. For example, a system initially deployed to allow employees easy access to a worksite might also be used later to track attendance, hours worked, or even movement at the worksite. Such concerns argue for clear documentation of both how the system will be used and protections to ensure that it will not be used for other, unacknowledged purposes.

More broadly, the social and cultural factors that influence willingness to participate in biometric systems run the gamut from trust in government and employers, to views about privacy and physical contact, to social involvement vs. isolation. Because the use of biometric systems depends on physical connections with individuals and because they are used for national security, law enforcement, social services, and so on, a host of societal issues should be addressed before they are deployed.

Facilitating Individual Participation

The adoption of biometric systems depends on the ease with which people can use them. In systems design it is critical to consider training in use of the system, ease of use (for example, are multiple steps, awkward actions, or complicated procedures required?), and management of errors (for example, how does the system recover from a mistake?). Designing usable systems also requires that the designers have some knowledge of the human users and operators, the context in which they will use the system, and their motivations and expectations.[2] Users of biometric sys-

---

[2]For more on usability and biometric systems, see "Usability and Biometrics: Ensuring Successful Biometric Systems." Available at http://zing.ncsl.nist.gov/biousa/docs/Usability_and_Biometrics_final2.pdf.

tems might include travelers, people whose first language is not English, employees of a particular company, shoppers, and so on. Contexts of use might include empty office buildings vs. busy airports, indoor vs. open air, lone individuals vs. groups, daily vs. semiannual, and so on. Motivations of operators might range from speeding up the check-in process to protecting personal information to preventing terrorism.

Discussions of usability tend to focus on narrow technical considerations such as the adequacy of the instructions for where and how to place the hand or finger to successfully engage with a biometric system (some of this was discussed briefly in the section "Operational Context" in Chapter 2). But broader considerations also affect usability. For example, providing a table where users can place their bags, purses, and other paraphernalia before interacting with the system may improve usability. Interfaces should take into account physical differences among people (height, girth, agility). If an interface is difficult for a particular person (tall or short, say) to use, then users are implicitly sorted into categories and may be treated differently for reasons unrelated to system goals.

Dealing with user diversity also leads to the challenge of providing user assistance. The presence of knowledgeable people providing help has been shown time and again to ease the pain of learning to use new systems and of managing errors. For a biometric system, it is important to have ways to mediate a variety of potential problems, ranging from individuals uncertain how to use the system to individuals who cannot present the trait needed (for example, if their fingerprints are hard to image). Earlier chapters considered failures to enroll and failures to acquire from technical and statistical perspectives, but how such failures are handled from the perspective of the user (who has "failed" in some sense) also has an impact on system effectiveness. If they are not handled carefully, some users may be less ready to participate, or even disenfranchised altogether; see below for the broader societal implications of disenfranchisement.

Expected frequency of use by a given individual is an important consideration in system design, since familiarity comes with repeated use. Typically, systems designed for infrequent use should be easy to learn, with readily interpreted instructions and help at hand, as usage procedures may not be remembered. On the other hand, people can learn to use even the most difficult systems if they have a chance to practice and learning is reinforced by frequent use and feedback. Thus, a system to be used by vacationers once or twice a year will have different requirements than one used daily by employees. While systems used frequently should avoid time-consuming operations to accomplish routine tasks such as entering a work area or logging on to the computer, they need not minimize initial training.

SOCIETAL IMPACT

The increasing use of biometric systems has broad social ramifications and one overarching consideration is proportionality. While the technical and engineering aspects of a system that contribute to its effectiveness are important, it is also useful to examine whether a proposed solution is proportional and appropriate to the problem it is aimed at solving.[3] Biometric systems' close connection to an individual, as described in the preceding section, means that even extremely effective technical solutions may turn out to be inappropriate due to perceived or actual side effects and means that proportionality—both how the system will be perceived in its user communities as well as possible side effects, even if the system is accurate and robust—must be considered when first examining the solution space. The rest of this section explores some of those potential side effects, including potential disenfranchisement of nonparticipants, privacy issues, and impact of varying cultural perspectives on individuality and identity.

Universality and Potential Disenfranchisement

Where biometric systems are used extensively, some members of the community may be deprived of their rights. Some individuals may not be able to enroll in a system or be recognized by it as a consequence of physical constraints, and still others may have characteristics that are not distinctive enough for the system to recognize. There will also be those who decline to participate on the basis of religious values, cultural norms, or even an aversion to the process. Religious beliefs about the body and sectarian jurisdiction over personal characteristics (for example, beards, headscarves) or interpersonal contact (for example, taking photographs, touching, exposing parts of the body) may make a biometric system an unacceptable intrusion. Mandatory or strongly encouraged use of such a system may undermine religious authority and create de facto discrimination against certain groups whose members are not allowed to travel freely, take certain jobs, or obtain certain services without violating their religious beliefs. Another category of people who may choose not to participate are those concerned about misuse or compromise of the system or its data—and its implications for privacy and personal liberty. Although a decision to participate or not may be an individual one, biometric sys-

---

[3]An example that received some media attention was a proposal to use fingerprint scanners to speed up a school lunch line. For reasons described elsewhere in this chapter, many in that community felt that such a use of the technology was disproportionate to the problem, regardless of its effectiveness.

tems can inadvertently affect groups whose shared characteristics make them less inclined to use the systems, assuming that participation is voluntary. Where use is mandatory—for example, in some military applications such as the U.S. Army's Biometric Automated Toolkit (BAT) system described in Appendix D—even more consideration of these issues may be needed.

Thus, while disenfranchisement in such cases may seem to affect only individuals, broad use of systems that are known to have these consequences can adversely affect the broader community if no appropriate relief is put in place. The community, including those not affected directly, may come to distrust the systems or the motivations of those deploying them. A system deployed in a community in which certain members are consistently unable to participate in the de facto methods of recognition without significant inconvenience may acquire an unwelcoming reputation no matter how benign the purposes for which it is deployed. Similar kinds of potential ostracism have been seen when formerly pedestrian-friendly or bus-friendly communities are transformed to focus on automobiles; those without driver's licenses or their own automobiles can become effectively disenfranchised in particular geographic locations.[4] Other technologies, such as the telephone and the Internet in the communications arena, have had similar effects.

Privacy as a Cultural Consideration

Biometric systems have the potential to collect and aggregate large amounts of information about individuals. Almost no popular discussion of biometric technologies and systems takes place without reference to privacy concerns, surveillance potential, and concerns about large databases of personal information being put to unknown uses. Privacy issues arise in a cultural context and have implications for individuals and society even apart from those that arise in legal and regulatory contexts. The problems arising from aggregating information records about individuals in various information systems and the potential for linking those records through a common identifier go well beyond biometrics, and the challenges raised have been addressed extensively elsewhere.

For example, a 2007 NRC report[5] that examined privacy in the digital age had a host of citations to other important work in this area. A thorough

---

[4]Langdon Winner, Do artifacts have politics? *Daedalus* 109(1) (1980), and How technology reweaves the fabric of society, *The Chronicle of Higher Education* 39(48) (1993). Donald A. MacKenzie and Judy Wajcman, eds., *The Social Shaping of Technology*, London: Open University Press (1985; 2nd ed., 1999).

[5]See NRC, *Engaging Privacy and Technology in a Digital Age*, Washington, D.C.: The National Academies Press (2007).

treatment of authentication technologies and privacy, with references to a host of sources, appears in the NRC report *Who Goes There? Authentication Through the Lens of Privacy* (2003), which treats the constitutional, statutory, and common law protections of privacy and their intersection with modern authentication technologies, including biometrics. A 2002 NRC report from the same project explored large-scale identity systems and potential technical and social challenges.[6] Almost all of the issues raised in these three NRC reports on technology and identity systems, with or without biometric components, also apply to biometric systems.[7] In addition, a recent symposium on privacy and the technologies of identity includes a series of scholarly papers on the subject. These papers refer to a wide range of sources.[8] This chapter does not seek to recapitulate this extensive literature, and instead briefly examines some ways biometric systems can contribute to the privacy challenges inherent in systems storing information about individuals.

Record Linkage and Compromise of Anonymity

Information of various kinds about individuals is routinely stored in a variety of databases. Linking such information—however imperfectly—in order to form profiles of individuals is also routinely done for purposes ranging from commercial marketing to law enforcement. The biometric data stored in information systems have the potential of becoming yet another avenue through which records within a system or across systems might be linked. This potential raises several questions: Under what circumstances is such linkage possible? If undesirable linkages are technically feasible, what technological and/or policy mechanisms would impede or prevent them? How could compliance with those mechanisms be monitored by those whose data are stored? What criteria should be used for deciding whether these mechanisms are needed? Depending on the anticipated uses of the personal data, policy and technical mechanisms may have to be put in place to prevent their unauthorized linking.

A challenge related to record linkage is the potential for erosion or compromise of anonymity. As discussed previously, in contrast to the wide choice of passwords available to an individual, there are a fairly limited number of biometric identifiers that a person can present, even when all possible combinations (for example, multiple fingers, face recognition

---

[6]See NRC, *IDs—Not That Easy: Questions About Nationwide Identity Systems*, Washington, D.C.: The National Academies Press (2002).

[7]See NRC, *Who Goes There? Authentication Through the Lens of Privacy*, Washington D.C.: The National Academies Press (2003).

[8]See Katherine Strandburg and Daniela Stan Raicu, eds., *Privacy and Technologies of Identity: A Cross-Disciplinary Conversation*, pp. 115-188 (2006).

coupled with hand geometry) are considered. Thus, even a biometric system that does not internally link an individual's biometric data with other identifying information may fail to preserve anonymity if it were to be linked using biometric data to another system that does connect biometric data to identity data. This means that even a well-designed biometric system with significant privacy and security protections may still compromise privacy when considered in a larger context. A related challenge is secondary use of data—that is, using data in ways other than originally specified or anticipated. The 2003 NRC report *Who Goes There?* examined secondary use in an authentication context. The challenge to privacy posed by secondary use of data in information systems generally, and particularly in data-intensive systems even without biometrics, is widely known.

Although it may seem that these concerns are specific to individuals, privacy considerations can have broad social effects beyond the individual,[9] as the discussion above on universality makes clear. Privacy breaches, however well-contained, can erode trust not only in the technological systems but also in the institutions that require their use. The potential for abuse of personal information can be sufficient to make certain segments of society reluctant to engage with particular technologies, systems, and institutions. Biometric systems carry their own particular challenges with respect to privacy in addition to many of those that have been identified for other information systems.

Covert Surveillance

Some recognition systems may function at a distance, making it possible to associate actions or data with a person without that person's explicit participation. Such tracking and collection of data has privacy implications not only for the person involved but for society as a whole. If these capabilities were to be broadly deployed, with their existence becoming broadly known and concern about their use becoming common, there would be potential distrust of the institutions that had deployed the technology. Even if knowledge of a capability is not widespread, the power that flows to those who control it may have unanticipated effects.

To date, widespread use of covert identification appears to be confined to movie plots. Contrary to popular belief, for example, the surveillance cameras used to investigate the 2005 London bombings did not perform biometric recognition as described in this report, because the cameras produced video searched by humans, not by machine. Nonetheless, several programs now pursuing recognition at a distance presage

---

[9]See, for instance, Priscilla Regan, *Legislating Privacy: Technology, Social Values, and Public Policy*, University of North Carolina Press Enduring Editions (2009).

such applications. While concerns about choice to participate are often dismissed by the biometrics industry, they must be addressed to take into account the target community's cultural values in order to gain acceptance and become broadly effective, especially as such systems become more pervasive or if covert biometric surveillance systems mature and become widely deployed.

Individuality and Identity

Because a biometric system recognizes the body, its applications may assume and embed particular Western notions of individuality and personal identity—namely, that the individual acts in self-interest and has autonomy over his or her actions. However in some non-Western contexts there are different views on the primacy of the individual, and more collectivist views of identity prevail. In these contexts agency and authority are not presumed to reside with individuals and autonomous action is not assumed. An individual's identity, in such cases, cannot easily be separated from that of the larger group of which the individual is a member. This undermines the assumptions of systems (biometric and otherwise) whose design expects that the actor is the individual (for example, accessing a bank account, doing a job, or making political decisions).[10]

Biometric systems are used to recognize individuals, but depending on the application and the cultural context, the broader system integrating biometric technologies may also need to recognize the position of individuals within the family, workplace, or community.[11] That is, an understanding of the relational dimensions of individual action whereby a person or group acts on behalf of another may be required. For example, within a workplace context an assistant may be required to take action on behalf another to check in for a flight or post personnel evaluations or in a community context a neighbor may be enlisted to pick up prescription medication for someone unable to do so.[12]

---

[10]There are microexamples of this sort of thing even within Western cultures. For example, administrative assistants are often given authority to make decisions and even sign documents for their bosses when their bosses are not present. Parents may act on behalf of their children, and spouses often are able to speak for each other.

[11]Note that a well-designed biometric system will integrate an appropriate notion of individual. The Walt Disney World entrance gate application of biometrics described in Appendix D creates an affinity group for the people entering the park using a set of tickets purchased at the same time. This association of the tickets to the group avoids unnecessary complication at the admission gate and reflects the common social context for ticket use without unduly weakening the value of biometric recognition.

[12]These examples are more properly accommodated by features in the broader system in which the biometrics components are integrated. The authorization system may allow delegation rather than recognize multiple individuals as the same entity.

Another way the identity of an individual can be conflated with that of a larger group is when biometric data are used for research. While this report urges extensive empirical research (of necessity on large groups of individuals) to achieve a stronger empirical foundation for biometric systems, such research is not without complications. When a group (such as an ethnic or racial group) is studied, whether for a medical, say, or a biometric purpose, the associated findings about that group can raise issues. Perhaps the individuals who were studied had not consented to the use of their individual data for research and the drawing of generalizations about their group, or perhaps consent was obtained but the results of the research are not welcomed by the individuals who participated. Another complication may arise when the results of a group study are made public and have an effect on individuals who are part of that group, whether or not they participated in the study. Even though individual enrollees in the database have given their consent, does the group qua group have the power to withhold consent for conduct of the research itself or publication of the findings? These issues apply beyond biometrics research, but biometric recognition's close association to individual bodies and notions of identity will inevitably heighten participants' sensitivity to the issues and necessitate that they be addressed with special care.

In addition to the identity issues raised by cultural considerations and role-based agency and the challenges of research on socially identifiable groups, biometric technologies explore the boundary between public and private information about an individual's body. The ability of these systems to categorize, monitor, and scrutinize persons through behavioral or biological characteristics raises the issue of the integrity of the *person*. The gathering of biometric data of all kinds (for example, fingerprint images, iris scans, brain scans, DNA, face imaging) that is associated with and defines the individual raises issues of the "informatized" body—a body that is represented not by human-observable anatomical and physical features but by the digital information about the body housed in databases. This has implications for how we ultimately perceive and conceive of the individual.[13]

---

[13]For more on this notion, see Irma van der Ploeg, 2007, "Genetics, biometrics and the informatization of the body," *Ann. Ist. Super. Sanità* 43(1): 44-50, and Emilio Mordini and Sonia Massar "Body biometrics and identity," *Bioethics* 22(9): 488-498 (2008). The former notes:

> The digital rendering of bodies allows forms of processing, of scrolling through, of datamining peoples' informational body in a way that resembles a bodily search. Beyond mere data privacy issues, integrity of the person, of the body itself is at stake here. Legal and ethical measures and protections should therefore perhaps be modelled analogous to bodily searches, and physical integrity issues. This issue is of particular relevance with regard to a curious aspect of this new body, namely that it has become *(re-)searchable at a distance*. The digitized body can be trans-

Biometric data contribute to new ways of knowing and defining persons as digitized information. This information is gathered during both routine and exceptional activities such as medical examinations, performance testing in sports, and users' interactions with biometric systems deployed in the various applications described elsewhere in this report. For some purposes, the observable physical body becomes less definitive and exclusive with respect to connoting who we are and is increasingly augmented (or even supplanted) by digital information about us. This information may ultimately figure into such decisions as who we date, mate, and hire and—conversely—who dates, mates, and hires us. The ability to recognize people by how they look or walk or talk is a human skill critical to social order and human survival. Some individuals are able to quickly recognize people they have not seen for years. Biometric systems that are able to perform these historically human acts of recognition at high rates of speed and on a massive scale may alter underlying assumptions about the uniqueness of these human capabilities and may blur previously clear boundaries between the human skills and social processes that control access to social spaces and bestow rights and duties, and the technological capabilities of biometric systems that recognize faces, gaits, and voices.

LEGAL ISSUES

Comprehensive discussion of legal issues associated with biometrics is well beyond the scope of this report. However, as with any scientific or technical issue, the assumptions made by engineers are very different from those made in the legal system. Understanding the broader context, including the legal context, within which biometric systems will operate, is important to achieving effectiveness. The use of biometrics brings with it important legal issues, especially the following: remediation, reliability, and, of course, privacy. Legal precedent on the use of biometrics technology is growing, with key cases stretching back decades,[14] and some recent

---

ported to places far removed, both in time and space, from the person belonging to the body concerned. Databases can be remotely accessed through network connections; they are built to save information and allow retrieval over extended periods of time. A bodily search or examination used to require the presence of the person involved—a premise so self-evident that to question it would be quite ridiculous. Moreover, this requirement rendered the idea of consenting to any bodily search at least a practicable possibility. Today, however, these matters are not so obvious any more.

[14]Cases include *U.S. v. Dionisio* (U.S. Supreme Court, 1973) and *Perkey v. Department of Motor Vehicles* (California Supreme Court, 1986).

cases[15] have raised serious questions as to the admissibility of biometric evidence in courts of law.

Remediation refers to the legal steps taken to deal with the fraudulent use of biometrics, such as identity fraud by altering or concealing biometric traits, altering biometric references, or using fake biometric samples to impersonate an individual. It also handles circumstances where individuals are incorrectly denied their due rights or access due to a false nonmatch. While earlier chapters examined the design and engineering implications of systems that should be able to (1) cope reasonably with such fraudulent attempts and implement security measures against them, and (2) gracefully handle individuals who are erroneously not matched through some secondary procedure, no system can be completely fraud- or error-proof. Thus, it will be important for policy and law to both address the perpetrator of identity fraud and induce system owners to create an environment that minimizes the opportunity for misuse of biometric samples—for example, by appropriately monitoring biometric sample presentation at points of enrollment and participation. It will also be important for policies to encourage appropriate and graceful management of false nonmatches. Reliability and privacy, and their potential intersections with biometric systems, are discussed below.

Reliability

Reliability has a social, as well as a technical, dimension. In the long run, biometric applications that make well-publicized or frequent errors will lose public support, even though some aspects of popular culture (such as police procedural television shows) have promoted the idea that forensic information, and in particular biometric data, is nearly infallible. Thus reliability, like privacy, is vital to the future of biometrics. Jurors relying on fingerprint evidence need assurance that they are convicting the right person. A security agency relying on voice analysis must feel confident that the person detained is a terrorist and not an innocent bystander. Context is crucial. Information may be reliable enough to begin an investigation, yet insufficiently trustworthy to send a person to prison. There is no guarantee that addressing reliability challenges alone would result in biometric systems becoming broadly accepted or even useful. In any particular setting, biometrics may or may not ensure the confidence needed for the system to be useful.

The reliability of biometric recognition has received considerable attention for many years. Numerous scholarly articles discuss whether particular forms of biometric evidence, such as DNA or fingerprints, are

---

[15]Such as *Maryland v. Rose* (Maryland Circuit Court, 2007).

admissible in court and whether they should be.[16] A recent NRC report took a broad look at forensics, not just biometric evidence, and concluded that "a body of research is required to establish the limits and measures of performance and to address the impact of sources of variability and potential bias."[17] After the misidentification and arrest of Brandon Mayfield in connection with the Madrid train bombings of 2004,[18] courts appear to be taking a cautious approach to biometric recognition, even though the error was ultimately attributed to human experts and the Department of Justice report does not fault the automated matching portion of the system. The reverse perception may set excessively high standards; that is, if the assumption is that all evidence must be up to the standards implied by certain popular culture phenomena, then cases in which resources were not available to meet those standards may face challenges. While conclusive studies on the effects of television crime drama on jurors have not been published[19] concerns over potential effects continue to surface in legal appeals.[20]

Without repeating the extensive studies mentioned above, it is helpful to consider how our society handles reliability concerns. A brief discussion of the admissibility of biometric evidence in court will be useful. As biometric systems are deployed more broadly and in more contexts, a public understanding of the extent to which the data they gather and results they produce can be relied upon will be critical. The discussion below explores some of the limitations and constraints that such data and results might face in a legal context. While biometric reliability is relevant

---

[16]See, for example, Julian Adams, Nuclear and mitochondrial DNA in the courtroom, *Journal of Law and Policy* 13:69 (2005); Sandy L. Zabell, Fingerprint evidence, *Journal of Law and Policy* 13:143 (2005). There have been substantive conferences dealing with these matters, such as the National Science Foundation Workshop on the Biometric Research Agenda (2003). The use of biometric evidence in court is a subset of the field of forensics; it concerns "the application of the natural and physical sciences to the resolution of conflicts within a legal context." See David L. Faigman et al., *Science in the Law: Forensic Science Issues*, West Group Publishing, p. 4 (2002). Also, forensic science "encompasses a broad range of disciplines, each with its own distinct practices." See National Research Council, *Strengthening Forensic Science in the United States: A Path Forward*, Washington, D.C.: The National Academies Press (2009), p. 38.

[17]National Research Council, *Strengthening Forensic Science in the United States: A Path Forward*, Washington, D.C.: The National Academies Press (2009). Also available at http://www.nap.edu/catalog.php?record_id=12589.

[18]Available at http://www.usdoj.gov/oig/special/s0601/PDF_list.htm; http://www.usdoj.gov/oig/special/s0601/exec.pdf.

[19]Tom R. Tyler, Is the CSI effect good science?, *Yale Law Journal* (The Pocket Part) (February 2006). Available at http://www.thepocketpart.org/2006/02/tyler.html.

[20]John Ellement, SJC chief decries influence of "CSI," *The Boston Globe*, December 11, 2009. Available at http://www.boston.com/news/local/massachusetts/articles/2009/12/11/mass_judge_says_theres_no_place_for_csi_in_real_courts/.

in numerous other settings, from border security to consumer transactions, the jury trial provides a highly visible insight into public views of reliability, and these views may well bear on other contexts.

The first point is that the strong consistency and accuracy sought by laboratory researchers and system developers is not always a goal of our legal system. While the Constitution provides broad norms binding the government, our system of federalism, in which substantial power is retained by the states, encourages diversity, even on fundamental issues. Even within a state, important decisions are made by local governments, individual judges, and juries. Given the changing nature of expert opinion, this sort of flexibility may not only be expected, it may also at times be welcomed.

Let us suppose that an individual in a given jurisdiction is charged with a crime, and the prosecutor seeks to introduce biometric evidence that is relevant to the guilt of that individual. Note that this biometric evidence may or may not be forensic evidence (for example, latent fingerprints). It might be a time-stamped log generated by a workplace entry biometric system. Or it might be information taken from the individual's PDA that has a fingerprint access mode. Typically, this evidence would be presented by an expert, someone who, unlike an ordinary witness, is allowed to give her opinion as to what the evidence shows. The prosecutor might, for example, want to use an expert on voice analysis who has studied an incriminating phone conversation with a novel biometric technique and wants to offer her opinion that the voice on the phone is most likely that of the defendant.

In our legal system the judge performs a gatekeeper function: He or she decides whether the expert testimony is sufficiently reliable to be presented to the jury. If the judge rules that it is, the expert witness testifies before the jury and is subject to cross-examination, as well as to rebuttal testimony by opposing experts. The jury is free to believe or disbelieve any or all of the experts. If the judge rules that the expert testimony is not sufficiently reliable, she cannot take the stand, and the prosecution's case might collapse.[21]

Frye and Daubert Standards

There are legal standards a judge uses in gatekeeping expert testimony. In many states, the judge uses the *Frye* standard, which asks

---

[21]This gatekeeper role for the judge represents a compromise. In theory, a jury might be allowed to hear anyone, and we would rely on cross-examination and opposing witnesses to assure accuracy. But our system is a bit more constrained than that. Some juries might believe the testimony of an astrologer, but no judge would admit it. On the other hand, our system does leave considerable power in the hands of the jury, since a judge's decision to admit expert testimony in no way guarantees that the jury will believe that testimony.

whether the expert opinion is based on a scientific technique that is generally accepted by the relevant scientific community.[22] In other states, and in the federal system, the judge uses the more recent *Daubert* approach, which requires that he balance a variety of factors, including whether a scientific approach has been tested and subjected to peer review and publication; its error rate; and its general acceptance as defined in *Frye*.[23] Later federal cases have applied the *Daubert* approach to all expert opinion, not just to scientific evidence,[24] and have held that appellate courts should reverse trial courts under *Daubert* only when they have abused their discretion.[25,26]

Numerous studies have compared *Frye* and *Daubert* and have reached varying conclusions as to which is the higher hurdle in various settings, although most would agree that the change from *Frye* to *Daubert* did not work a sea change in admissibility decisions.[27] The use of lay juries and judges who are not scientists means decisions on the admissibility of evidence will reflect social values more than technical expertise. For example, despite considerable controversy over its reliability, eyewitness testimony is routinely allowed in court; indeed, it is not always possible to introduce expert opinion to cast doubt on that testimony.[28] A person may not have very good vision or a good memory, but may be allowed to testify before a jury about what he or she saw.

Why do we insist on a gatekeeper for technical experts but not for eyewitnesses? There is apparently a belief that a jury that has a common-sense understanding of the abilities of an eyewitness may defer too much to a highly credentialed expert. If that expert is articulate and persuasive, the usual checks of cross-examination and rival expertise may not be adequate. So the judge acts as a gatekeeper, using *Frye* or *Daubert* to keep from the jury theories that lack adequate signs of objective reliability.

Of course, when an expert witness testifies and a rival expert is called to rebut, the ensuing battle of the experts may not be helpful for the jury or for society. In a survey conducted by the Federal Judicial Center, judges frequently complained about experts who "abandon objectivity

---

[22]*Frye v. United States*, 54 App. D.C. 46, 47, 293 F. 1013, 1014 (1923).

[23]*Daubert v. Merrell Dow Pharmaceuticals, Inc.*, 509 U.S. 579 (1993).

[24]*Kumho Tire Company, Ltd. v. Carmichael*, 526 U.S. 137 (1999).

[25]*General Electric v. Joiner*, 522 U.S. 136 (1997).

[26]For an analysis of which states use *Frye*, which use the full-blown federal approach of *Daubert*, *Kumho*, and *Joiner*, and which use *Daubert* alone, see David Bernstein and Jeffrey Jackson, The Daubert trilogy in the states, *Jurimetrics Journal* 44(351).

[27]David A. Sklansky, *Evidence: Cases, Commentary, and Problems* (2003), Aspen Publishers, pp. 468-470 (cites numerous studies).

[28]See, for example, *United States v. Smithers*, 212 F. 3rd 306 (6th Circuit 2000).

and become advocates for the side that hired them."[29] One possible solution, which has been promoted in the federal system by United States Supreme Court Justice Stephen Breyer, is to have the judge call an expert drawn from a list provided by leading scientific organizations.[30]

The upshot of this for biometric systems and technologies is that the admissibility of biometric evidence will not be consistent throughout the country. States may differ on what is reliable, and even judges in a given state may sometimes differ. On the other hand, over time, a consensus will emerge in some areas. Polygraph evidence, for example, is generally denied admission in court[31] while DNA evidence is generally admitted.[32]

Finally, the legal system will likely never render a verdict on biometric evidence as a whole. As one would expect, and hope, each type of evidence will be evaluated on its own when a judge exercises his or her gatekeeper function and when a jury performs its ultimate decision-making role. In these legal settings at least, societal attitudes to biometrics are much less important than societal attitudes toward specific biometric techniques. It is possible that opinions about techniques may change dramatically over time. For example, fingerprints are familiar and might be generally viewed as trustworthy, but people might be more skeptical of voiceprints or another more esoteric technology, no matter what the opinion of experts.

Privacy in a Legal Context and Potential Implications for Biometrics

Virtually every discussion of the social implications of biometrics begins with privacy, and for good reason. Biometric information is part of an individual's identity (in the colloquial sense of the term), and a loss of control over that information can threaten autonomy and liberty.

In thinking about these concerns, context is crucial. There is no one-size-fits-all set of rules to ensure that a biometric system adequately protects privacy. It will be necessary instead to formulate policies for specific situations and to evaluate them as time passes. Nor is there a guarantee of success. Biometric technologies do not inevitably threaten privacy: They could be neutral in that regard or they could even enhance privacy. The outcome depends on the choices our society makes.

---

[29]Molly Treadwell Johnson et al., "Expert Testimony in Federal Civil Trials: A Preliminary Analysis," Federal Judicial Center (2000), at 5-6. Available at http://www.fjc.gov/public/pdf.nsf/lookup/ExpTesti.pdf/$file/ExpTesti.pdf.

[30]See *General Electric v. Joiner*, 522 U.S. 136 (1997) at 149 (Breyer, J., concurring).

[31]See, for example, *State v. Porter*, 698 A.2d 739 (Connecticut, 1997).

[32]Julian Adams, Nuclear and mitochondrial DNA in the courtroom, *Journal of Law and Policy* 13:69 (2005).

The Public Sector and Privacy Rights

Privacy implicates a variety of constitutional norms in our legal culture. In the public sector, the government's failure to safeguard personal information can implicate due process, its abuse of information can stifle free speech, and its failure to have an adequate basis for acquiring information can challenge our protections against unreasonable search and seizure as well as self-incrimination.

Rather than reiterate the many surveys that have come before, here we consider two recent decisions of the Supreme Court that relate to the intersection of biometrics and privacy rights. While these cases do not directly concern biometrics, some of their potential implications for that field are noted.

Hiibel v. Sixth Judicial District Court   Consider first the Supreme Court's 2004 decision in *Hiibel v. Sixth Judicial District Court*.[33] Deputy Sheriff Lee Dove of Humboldt County, Nevada, was dispatched to investigate a telephone call reporting that a man in a red and silver GMC truck on Grass Valley Road had assaulted a woman. When Dove found the truck, he saw a woman sitting in it and a man standing beside it. Dove asked the man for identification, and the man refused. As Dove repeated the request, the man became agitated and said he had done nothing wrong, to which Dove replied that he wanted to know who the man was and why he was there. Eventually, after requesting identification 11 times and warning the man he would be arrested for failure to comply, Dove arrested the individual.

The arrested individual, Larry Dudley Hiibel, was charged under Nevada law with having obstructed a police officer by failing to identify himself. Nevada and some other states make it a crime to fail to identify oneself when stopped by a law enforcement officer. Hiibel was convicted and fined $250. His appeal, contending the Nevada law violated his Fourth and Fifth Amendment rights, ultimately reached the Supreme Court. A five-member majority of the Court affirmed Hiibel's conviction. Four Justices dissented. For more detail on the arguments and decisions, see Box 4.1.

Biometrics makes the issues in *Hiibel* more pressing than ever. Consider police use of facial recognition technology, which has been tried on a limited basis in a few jurisdictions.[34] Let us imagine a full-blown system in which officers on patrol have the ability to take digital images and compare them in real time with a variety of databases. Suppose Officer Dove

---

[33]542 U.S. 177 (2004).

[34]See, for example, Outsmarting the bad guys, *Los Angeles Times*, September 29, 2005, p. B2.

BOX 4.1
The Legal Arguments in *Hiibel*

Nevada recognized that Officer Dove lacked the probable cause required by the Fourth Amendment to arrest Hiibel. It relied on the Supreme Court's 1968 opinion, which held that "reasonable suspicion" was an adequate Fourth Amendment standard when officers simply stop individuals against their will and pat them down to make sure they are not carrying a weapon (*Terry v. Ohio*, 392 U.S. 1 (1968)). Under *Terry* and later decisions, a suspicious officer can stop an individual and ask questions (also *INS v. Delgado*, 466 U.S. 210 (1984)). The *Terry* decision left open what recourse, if any, a police officer had if the individual refused to answer those questions. In a concurring opinion in *Terry* (392 U.S., at 34), Justice White wrote as follows: "Of course, the person stopped is not obliged to answer, answers may not be compelled, and refusal to answer furnishes no basis for arrest, although it may alert the officer to the need for continued observation."

Hiibel urged the Court to adopt Justice White's view. The Fourth Amendment's protection of privacy, in Hiibel's view, should prevent the government from forcing him to reveal his identity or answer other questions in the absence of probable cause that he had done something wrong. The closely related Fifth Amendment protection against self-incrimination should have the same result.

Justice Kennedy's opinion for the Court rejected Hiibel's arguments and thus Justice White's assumption in *Terry* that when you are stopped by a police officer without probable cause you do not have to answer questions. The Court reasoned that the state of Nevada had a strong interest in requiring people to identify themselves (542 U.S., at 186):

> Knowledge of identity may inform an officer that a suspect is wanted for another offense, or has a record of violence or mental disorder. On the other hand, knowing identity may help clear a suspect and allow the police to concentrate their efforts elsewhere. Identity

---

had not asked Larry Hiibel a single question—that instead, upon finding the GMC truck with a woman inside and a man beside it, Dove, without the man's consent, had taken a digital image and compared it with images in a variety of databases. Would Hiibel have suffered an unacceptable invasion of privacy?

Much like the question of whether the failure to answer a police officer's questions may be criminalized, this is a question for the legislature. Thus states (and the federal government) would have to decide whether to authorize their law enforcement officials to use facial recognition technology in this way.

In states that do allow the technology, adversely affected individuals could still go to court and argue that their constitutional rights had been violated. The narrowly drawn, five-to-four decision in *Hiibel* does not resolve the question. In one sense, the government is in a stronger position here. The self-incrimination issue is not present with facial recognition technology, since a Fifth Amendment claim requires that you be com-

may prove particularly important in cases such as this, where the police are investigating what appears to be a domestic assault. Officers called to investigate domestic disputes need to know whom they are dealing with in order to assess the situation, the threat to their own safety, and possible danger to the potential victim.

Justice Kennedy's opinion for the Court also rejected Hiibel's Fifth Amendment claim, although it left open the possibility that such a claim might succeed in a different situation. The Court reasoned that Hiibel had never explained how revealing his name would incriminate him. "As best we can tell," the Court said, "petitioner refused to identify himself only because he thought his name was none of the officer's business" (idem, at 190). The Court noted that "a case may arise where there is a substantial allegation that furnishing identity at the time of a stop would have given the police a link in the chain of evidence needed to convict the individual of a separate offense" (idem, at 191). In such a case, the Court would revisit the Fifth Amendment issue.

The four dissenting justices believed that Justice White had been correct in Terry: Given the absence of probable cause, the state cannot invade one's privacy by compelling an answer to its questions. One of the dissenters, Justice Stevens, emphasized that the Nevada statute imposed a duty to speak on a "specific class of individuals"—namely, those who had been detained by the police (idem, at 191, Stevens, dissenting). Stevens characterized the privacy interest of that group in the following terms (idem at 196, Stevens, dissenting):

> A person's identity obviously bears informational and incriminating worth, 'even if the [name] itself is not inculpatory.' . . . A name can provide the key to a broad array of information about the person, particularly in the hands of a police officer with access to a range of law enforcement databases. And that information, in turn, can be tremendously useful in a criminal prosecution. It is therefore quite wrong to suggest that a person's identity provides a link in the chain to incriminating evidence 'only in unusual circumstances.'

pelled to testify. While the Court in *Hiibel* assumed that giving one's name is testimonial, it had previously held that the provision of nontestimonial evidence, such as a blood sample or a fingerprint, is not.[35] However, the closely related Fourth Amendment issue remains: Does taking the image of an individual who is not subject to arrest without his or her consent and comparing it to a database constitute reasonable search and seizure?

Of course, a police officer has always been free to look at any individual and compare his likeness with that on wanted posters. But to many Americans, and potentially many legislatures, the use of technology to help do this makes a difference (see the discussion earlier in this chapter of individuals and identity). The possible loss of privacy posed by automated facial recognition technology may or may not be outweighed by possibly better law enforcement.

---

[35]See *Schmerber v. California*, 384 U.S. 757 (1966).

The opinions in *Hiibel* illuminate the competing considerations. To some, Justice Kennedy's image of police officers who can quickly separate those who are dangerous or wanted from those who are at risk or innocent is an attractive picture indeed. Others are more likely to see Justice Stevens's world, where facial recognition databases cast a wide net to ensnare people stopped without probable cause.

The reliability of a biometric system, such as the facial recognition system hypothesized here, is obviously relevant to this inquiry. No one wants to be arrested for a crime he did not commit, and no one wants dangerous felons to go free because of devices that malfunction. But even a reliable system does not function successfully without additional safeguards. Lurking behind Justice Stevens's image of law enforcement databases is a concern that technology makes it easier than ever to invade privacy by allowing information gathered for one purpose to be used for another.

This, of course, is an important concern for those who fear government databases might be improperly used for political purposes. The constitutional issue here is typically framed in terms of our First Amendment rights to free speech and assembly. The Supreme Court has often found that these rights include the right to remain anonymous in certain settings.[36]

In 1999 in *Buckley v. American Constitutional Law Foundation*,[37] the Court struck down a Colorado requirement that individuals circulating ballot initiative petitions must wear a badge bearing their name. The Court noted here that ballot initiatives were often very controversial and that the requirement deterred participation in the political process, in effect saying that in some settings, facial recognition technology could compromise free speech by in essence making everyone wear an identification badge.

Of course, we give up some measure of privacy when we appear in public. And our remaining privacy interests are not absolute. But judicial recognitions of the importance of anonymity give us a way to understand

---

[36]In 1958, the Court stuck down an Alabama statute requiring organizations to disclose their membership to the state. The plaintiff was the NAACP, which clearly would have been threatened by the state's requirement (see *NAACP v. Alabama*, 357 U.S. 449 (1958)). Since then, the Court has invalidated a Los Angeles ordinance requiring any publicly distributed handbill to identify its author, as well as a broader ban in Ohio on anonymous campaign literature (see *Talley v. California*, 362 U.S. 60 (1960); *McIntyre v. Ohio Elections Commission*, 514 U.S. 334 (1995)). These results and others like them are hardly surprising, given our political history: The Federalist Papers, which built support for the Constitution and which play a role in its interpretation to this day, were published under pseudonyms.

[37]525 U.S. 182 (1999).

Larry Hiibel's insistence that he did not identify himself because his name was "none of the officer's business."[38]

It bears repeating that there is nothing intrinsic to biometrics that automatically aggravates Larry Hiibel's problem. Police officers relying on their own judgment unaided by technology can easily make politically or racially motivated decisions that improperly invade privacy. In some settings, biometric systems could alleviate rather than worsen these problems, by relieving the individual law enforcement agent of the recognition task and assigning it to a common automated system that presumably performs equally and repeatedly for all agents. Such a system would, of course, be subject to all of the usual technological and environmental factors discussed elsewhere in this report that might degrade its effectiveness.

In the end, many aspects of the intersection between the *Hiibel* decision and biometrics come down to control over the uses of databases. Consider again a facial recognition system employed by an officer on patrol. A legislature might understandably want to reap the benefits of such a system while eliminating its misuse. Suppose the state curtailed the use of the system by saying that images of those under suspicion should be compared only to a database of convicted felons. Is there any constitutional requirement that this limit be abided by? The Supreme Court has not answered this question, although it has suggested that a state might violate due process if it does not take reasonable steps to stop unwarranted disclosures of data.

The suggestion came in the Court's decision in *Whalen v. Roe*, 429 U.S. 589 (1977). A New York statute required that prescriptions for legitimate but addictive drugs be recorded on a computer database to prevent abuses such as users obtaining prescriptions from more than one doctor. Although the computer system was set up to prevent leaks and public disclosure of the identity of patients was made a crime, the system was challenged by those who feared that the information could get out, stigmatizing patients as addicts in violation of their privacy rights. The statute was upheld, with the Court noting that no evidence of information falling into the wrong hands had been presented.[39] In his opinion for the Court, Justice Stevens, in an oft-quoted passage (idem at 605-606), left open the possibility that some future database might not be constitutionally acceptable if it were not adequately protected against improper use:

> We are not unaware of the threat to privacy implicit in the accumulation of vast amounts of personal information in computerized data banks or other massive government files. . . . The right to collect and use such data

---

[38]*Hiibel v. Sixth Judicial District Court*, 542 U.S. 177, 190 (2004).
[39]429 U.S., at 601.

for public purposes is typically accompanied by a concomitant statutory or regulatory duty to avoid unwarranted disclosures. Recognizing that in some circumstances that duty arguably has its roots in the Constitution, nevertheless New York's statutory scheme . . . evidences a proper concern with, and protection of, the individual's interest in privacy. We therefore need not, and do not, decide any question which might be presented by the unwarranted disclosure of accumulated private data whether intentional or unintentional or by a system that did not contain comparable security provisions. We simply hold that this record does not establish an invasion of any right or liberty protected by the Fourteenth Amendment.

This recognition that the due process clause of the Fourteenth Amendment might protect informational privacy is clearly important. It suggests that a government biometric database with inadequate safeguards could be successfully challenged by an individual in that database on the ground that the government had violated his or her liberty. Moreover, regardless of whether a challenge in court would succeed, it is clear that the public desires protection from unwarranted disclosures from databases of all kinds. Biometric systems will be judged by that standard.

Kyllo v. United States   The other recent Supreme Court decision that casts light on privacy and biometrics is *Kyllo v. United States.*[40] Federal Agent William Elliott suspected that marijuana was being grown in the home of Danny Kyllo, but he lacked the probable cause necessary to obtain a search warrant. Accordingly, Agent Elliott sat in a car in the street next to Kyllo's home and used a thermal imager to scan the residence. The imager detected infrared radiation coming from Kyllo's house. The pattern revealed that portions of the house were hotter than the rest of the house and the neighboring homes. Agent Elliott concluded that Kyllo was using halide lights to grow marijuana. Using this and other information, Elliott obtained a warrant for a search. Once inside Kyllo's home, federal agents found more than a hundred marijuana plants. Kyllo, who was eventually convicted of manufacturing marijuana, appealed on the ground that using a thermal imager without probable cause constituted an unreasonable search under the Fourth Amendment. The Supreme Court, in another 5-4 decision, agreed with Kyllo. See Box 4.2 for an overview of the arguments used.

Two main lessons for biometrics emerge from *Kyllo*. The first is that as a technology advances and becomes widespread, our zone of constitutionally guaranteed privacy shrinks. The Court recognizes, moreover, that our society's expectations of privacy set the baseline for laws

---

[40]533 U.S. 27 (2001).

governing searches. In *California v. Ciraolo*, noted in Box 4.2, the Court, in upholding aerial surveillance of a fenced backyard, explicitly said that "in an age where private and commercial flight in the public airways is routine, it is unreasonable to expect" that one's backyard is private.[41] In *Kyllo* the Court implied that if thermal imagers had been in common use its decision would have been different. Thus if the day comes when a biometric device that analyzes voices can function from a hundred feet or so and extend its range into a private home, and if use of that device becomes widespread, the Fourth Amendment will have little application.

If the first lesson offers a boost for biometrics, the second lesson does the opposite. All nine justices in *Kyllo* expressed concern about future technologies that impinge on privacy. The majority wanted to step in now; the dissent wanted to let the legislatures have the first crack at controlling future developments. On this question, the nine justices on the Supreme Court are representative of a wide span of public opinion. The ability of the government to use biometrics to, say, track people's movements around their neighborhood or even inside their own homes will raise red flags for those concerned about privacy and government intrusion. Public assent, so crucial, is likely to be lacking unless the application of biometrics is highly justifiable and carefully circumscribed.

Finally, going from protections provided by the Constitution to the less lofty ones provided by statute, we find a series of federal and state laws that tend to control certain government databases on a sector-by-sector basis with varying success. Many of these statutes incorporate the Fair Information Practices code developed by the then-Department of Health, Education and Welfare in 1973 and incorporated into the federal Privacy Act of 1974. The strengths and weaknesses of these statutes have been exhaustively analyzed.[42] As *Hiibel*, *Kyllo*, and the other cases mentioned above demonstrate, in many sensitive areas an individual has no statutory protections and must rely on constitutional arguments.

The Private Sector and Privacy Rights

In *Whalen v. Roe*, Justice Stevens spoke of the threat to privacy posed by "massive government files."[43] But what about information held in private hands? Many Americans are just as concerned or even more so about the loss of privacy when personal information is given to their employer or demanded in an online private transaction. Suppose a bank gives its

---

[41]476 U.S., at 215.

[42]See, for example, Daniel J. Solove, Marc Rotenberg, and Paul M. Schwartz, *Information Privacy Law*, 2nd edition, pp. 523-622, New York: Aspen Publishers (2006).

[43]429 U.S., at 605.

BOX 4.2
The Legal Arguments in *Kyllo*

Although Justice Scalia's opinion for the Court emphasized the Constitution's traditional protection for the privacy of the home, he recognized that visual surveillance of the home without probable cause has long been allowed. In addition, his opinion for the Court conceded that "it would be foolish to contend that the degree of privacy secured to citizens by the Fourth Amendment has been entirely unaffected by the advance of technology" (533 U.S., at 33-34). For example, the Court had permitted aerial surveillance of the backyard of a private house, even though a fence shielded the yard from the street (*California v. Ciraolo*, 476 U.S. 207 (1986)).

Why did the Court disallow the use of the evidence adduced by the thermal imager? The Court relied on a test that stemmed from the case of *Katz v. the United States* (389 U.S. 347 (1967)). *Katz* upheld a challenge to warrantless eavesdropping by an electronic listening device placed on the outside of a telephone booth because Katz had justifiably relied on the privacy of the booth. The test held that "a Fourth Amendment search occurs when the government violates a subjective expectation of privacy that society recognizes as reasonable" (idem, at 33).

The Court recognized that the *Katz* test had been criticized as circular and unpredictable (idem, at 34), but it declined to revisit *Katz* in the setting of the thermal imaging of a private home. The Court also recognized that changing societal expectations of privacy affect Fourth Amendment rights under the *Katz* approach. In its opinion, the Court twice noted that the search of Kyllo's home was being set aside in part because the thermal imaging device was "not in general public use" (idem, at 34 and 40).

A final feature of the majority's opinion was the evident concern that more advanced technologies could reach into the privacy of the home. The Court said that "while the technology used in the present case was relatively crude, the rule we

---

customers the option of using fingerprints rather than a password to access an ATM. The bank may believe it is enhancing privacy because a password is easily stolen. But if the bank's fingerprint database is not adequately secured, a purposeful or accidental disclosure of that data could lead to identity theft. Suppose an employer conducts a biometric scan of its workers to facilitate access to the secure workplace when a badge is lost. If the biometric modality chosen happened to also reveal information about a worker's health, that information could be misused by the employer, by insurance companies, or others.

Justice Stevens's reference to "government" files was not inadvertent. The individual freedoms guaranteed by the U.S. Constitution are virtually all protections against government overreach. The Bill of Rights protects us from government suppression of free speech and religion, government establishment of religion, improper searches by government officials, deprivations of due process by the government, and so on. Similarly,

adopt must take account of more sophisticated systems that are already in use or in development" (p. 36). The Court then detailed what it had in mind (idem, at 36, note 3):

> The ability to "see" through walls and other opaque barriers is a clear, and scientifically feasible, goal of law enforcement research and development. The National Law Enforcement and Corrections Technology Center, a program within the United States Department of Justice, features on its Internet Website projects that include a "Radar-Based Through-the-Wall Surveillance System," "Handheld Ultrasound Through the Wall Surveillance," and a "Radar Flashlight" that "will enable law enforcement officers to detect individuals through interior building walls."

The four dissenting justices believed that Danny Kyllo had no reasonable expectation of privacy in heat emissions that were being sensed after they had left his house. In the dissenters' view, Agent Elliott's use of a "fairly primitive thermal imager" was no different than if he had noticed that Kyllo's house was warmer than a nearby building because "snow melts at different rates across its surfaces" (pp. 41 and 43, Stevens, dissenting). But the dissent was not prepared to say that advanced technology should always be absolved from Fourth Amendment scrutiny because it is nothing more than an enhancement of our senses. On that subject, Justice Stevens's dissent on p. 51 took a wait-and-see attitude:

> Although the Court is properly and commendably concerned about the threats to privacy that may flow from advances in the technology available to the law enforcement profession, it has unfortunately failed to heed the tried and true counsel of judicial restraint. Instead of concentrating on the rather mundane issue that is actually presented by the case before it, the Court has endeavored to craft an all-encompassing rule for the future. It would be far wiser to give legislators an unimpeded opportunity to grapple with these emerging issues rather than to shackle them with prematurely devised constitutional constraints.

Articles I, II, and III of the Constitution protect us from oppression by dividing government power between the federal and state governments and by dividing the federal government's power among the legislative, executive, and judicial branches.

The infringement of privacy by a private entity, including privacy of biometric information, can be protected against by legislation. But some limitations to this approach should be noted at the outset. While federal laws can be drafted that preempt state action, most such laws leave room for complementary state regulation, leading to debates over coverage. Moreover, when the federal government does not preempt or when it is silent, state laws typically differ from state to state, raising problems for businesses seeking to comply with the law and for enforcement efforts. Finally, only constitutional protections extend to minorities who lose out in legislative battles. For example, if the private sector depriving you of a job is in compliance with all relevant legislation, you have no recourse in court even if you believe you have been treated unfairly.[44]

Biometric information held by private companies is subject to constraints beyond those imposed by legislation. One such constraint might be self-regulation, which a company might impose to gain a market edge. Another such constraint might be common-law protections. An employee or a consumer might enter into a contract with a company that promises to protect biometric data and would risk breach of contract if it did not do so. Similarly, a company that failed to meet accepted standards in protecting information might be liable for negligence in a tort suit. However, continuing public concern about privacy suggests that market failures and the limits of relief achievable with retrospective common law make further legislative action likely.

There are many possible ways to regulate biometric technologies and systems that might provide needed protections for the public and build its confidence in the private sector. Some would be relevant to government databases as well. In response to the continuing concern over identity theft and fraud, some jurisdictions are considering enacting laws to prohibit the selling and sharing of an individual's biometric data, absent consent or compelling circumstances.[45]

Privacy concerns should be attended to when a biometric deployment is being implemented. For example, in answering the question of whether to store biometric data as processed references or in source samples or images, which approach best protects privacy should be considered. The same is true of the choice between local or centralized storage of biometric data—a choice that has significant security and privacy implications. Encryption of biometric data is often vital. Perhaps most important, the use of biometric systems should be defined and limited at the outset of a program, by legislation when appropriate. The temptation to use information for new purposes never justified to the public should be resisted.

In the end, working as hard on privacy as on technical success will help assure that biometric programs maintain or even enhance individual autonomy as they achieve social goals. Failure to do so will result in biometric programs that undermine American values while potentially bringing about their own failure due to public resistance.

---

[44]The NRC report *Who Goes There? Authentication Through the Lens of Privacy* (2003) found that personal information held by the private sector is afforded weaker statutory protections than information held by the federal or state governments and that much detailed personal information in the hands of businesses is available for reuse and resale to private third parties or to the government, with little in the way of legal standards or procedural protections.

[45]Illinois passed such a law in 2008 (740 ILCS 14/), the Biometric Information Privacy Act. Available at http://www.ilga.gov/.

DATA POLICIES

Biometric data are personally identifying information.[46] Thus biometric systems have the potential to collect not only pattern recognition information captured by sensors, but also other information that can be associated with the biometric data themselves or with data records already contained within the system. Depending on the biometric system, this information could include time and location of use, identification data (for example, name or Social Security number, and so on) and, in some cases, medical measurements (for example, glucose levels).[47] Additional data may be created when a decision is generated by the system (positive or negative recognition) that may be stored or shared with another system. Given the increasing volumes and kinds of data associated with a biometric system, data policies are important to answer a variety of questions that arise regarding sharing, storage, integrity, and confidentiality of the biometric system data.

Biometric systems are often associated with an identity system. Biometric data may be correlated across identity systems to recognize individuals. The data associated with an individual collected by different organizations using the same biometric modality may be similar but almost certainly not identical, because the sample acquisition for enrollment will vary (see Chapter 1 for elaboration on sources of uncertainty and variation in biometric systems).

An earlier NRC report addressed a set of questions and issues that arise particularly in the context of an identity system. For the most part, they apply to biometric recognition systems. The questions are reprinted for reference.[48]

- What is the purpose of the system? Possible purposes of an identity system include expediting and/or tracking travel; prospectively monitoring individuals' activities in order to detect suspicious acts; retrospectively identifying perpetrators of crimes.
- What is the scope of the population to whom an "ID" would be issued and, presumably, recorded in the system? How would the identities of these individuals be authenticated?

---

[46]The Data Protection Working Party is the independent European Union advisory body on data protection and privacy, established under Article 29 of Directive 95/46/EC. It determined that in most cases biometric data are personal data and can in all cases be considered as "information relating to a natural person." Available at http://ec.europa.eu/justice_home/fsj/privacy/docs/wpdocs/2003/wp80_en.pdf.

[47]While biometric data captured by systems using the most common modalities do not contain medical information, some emerging technologies capture traits such as heartbeat patterns, which directly convey medical data.

[48]NRC, *IDs—Not That Easy: Questions About Nationwide Identity Systems.* Washington, D.C.: The National Academies Press, pp. 9-11 (2002).

• What is the scope of the data that would be gathered about in-dividuals participating in the system and correlated with their system identity? "Identification systems," despite the name, often do much more than just identify individuals; many identity systems use IDs as keys to a much larger collection of data. Are these data identity data only (and what is meant by identity data)? Or are other data collected, stored, and/or analyzed as well? With what confidence would the accuracy and quality of this data be established and subsequently determined?

• Who would be the user(s) of the system (as opposed to those who would participate in the system by having an ID)? If the public sector or government will be the primary user, what parts of the government will be users, in what contexts, and with what constraints? In what setting(s) in the public sphere would such a system be used? Would state and lo-cal governments have access to the system? Would the private sector be allowed to use the system? What entities in the private sector would be allowed to use the system? Who could contribute, view, and/or edit data in the system?

• What types of use would be allowed? Who would be able to ask for an ID, and under what circumstances? Assuming that there are datasets associated with an individual's identity, what types of queries would be permitted (e.g., "Is this person allowed to travel?" "Does this person have a criminal record?"). Beyond simple queries, would analysis and data mining of the information collected be permitted? If so, who would be allowed to do such analysis and for what purpose(s)?

• Would participation in and/or identification by the system be vol-untary or mandatory? In addition, would participants have to be aware of or consent to having their IDs checked (as opposed to, for example, being subjected to surreptitious facial recognition)?

• What legal structures protect the system's integrity as well as the data subject's privacy and due process rights, and which structures de-termine the liability of the government and relying parties for system misuse or failure?

Information-Sharing Issues

With increased use of biometrics, there is legitimate concern about how information stored in biometric databases might be shared. Sharing can extend the administrative reach of biometric findings. It could also afford valuable facts for management and research studies. However, such sharing presents significant privacy and confidentiality challenges. Systematic approaches in law, regulation, or technology to resolve the tension between the evident demand to share more biometric information and the cautions of privacy and technology are lacking. In particular, no comprehensive federal policy exists to guide sharing information from biometric databases.[49]

The sharing of information from biometric databases occurs when the records in one database are integrated with those in other databases and when data are disseminated directly to users. Most biometric systems today involve databases to which biometric samples captured from a population of individuals have been submitted and are later searched to find matching enrolled individuals with the same biometric characteristics. In some systems, a sample is compared to the reference biometric data associated with the claimed identity (which means maintaining a database of reference data, even if it is not searched at each use). A national system of ID cards or passports based on biometric data would rely on a database.

The costs of electronically capturing biometric samples and storing the data continue to drop, as do the costs of data integration and dissemination, and the technical ability to do so is expanding as well, serving to increase interest in storing biometric information in databases. The benefits of storage capacity and data integration and sharing could include the following:

• *Administrative efficiencies.* One of the many applications might be giving a homeless shelter the ability to check whether an applicant has a criminal record; another might be allowing law enforcement to check whether a particular individual used a certain facility at a certain time.

• *Business purposes.* Sports teams could collaborate to examine the usage patterns and demographics of their season ticket holders, perhaps to avoid issuing more tickets than they have seats or to look for joint advertising opportunities.

• *Research uses.* Biometric data are needed for testing biometric system performance and for developing new systems and features.[50]

Although processing and sharing biometric information can bring many benefits, there are also concerns that stem from the ease with which biometrics technology integrates with database technology, increasing the likelihood of privacy violations. For this reason, it has been suggested

---

[49]In addition to the lack of policy guidance, there can be logistical and practical challenges to information sharing. For example, even within the DOD structure, the GAO has found gaps in biometrics sharing across missions. See http://www.gao.gov/products/GAO-09-49. For a more thorough treatment of information sharing and privacy, not just with respect to biometrics but also generally, see Peter Swire, Privacy and information sharing in the war on terrorism, *Villanova Law Review* 51:260 (2006).

[50]A. Ross, S. Crihalmeanu, L. Hornak, and S. Schuckers, A centralized web-enabled multimodal biometric database, *Proceedings of the 2004 Biometric Consortium Conference* (BCC), September 2004.

that privacy must be designed into the systems rather than added on at a later time.[51]

Currently there does not seem to be much of an ingrained culture of privacy protection for biometric databases, beyond that which exists for information systems generally or state and local efforts such as the Illinois statute mentioned previously. With the exception of some agencies, mainly statistical agencies, there is little historical tradition of maintaining the confidentiality of biometric databases. This is in spite of the fact that if biometric data associated with an individual falls into the wrong hands that individual could be at risk of identity theft.[52] Moreover, sharing of information from biometric databases raises questions of (1) whether the information would be used for purposes not intended or inconsistent with the purposes of the original biometric application and (2) what information about intended uses of the system should be disclosed to users and how that information should be presented.

Protection of Biometric Data

The protection of personal information is not the only reason for protecting biometric data. Another is the desire to prevent third parties from linking records between systems, determining the enrolled users in a system, or discovering a doppelganger (an individual who is a close match for an enrolled user). The encryption of biometric data stored in centralized databases or on a personal device such as a smart card, coupled with appropriate security measures to limit probing of the database, can be effective in countering these threats. Encryption and database protection, however, are insufficient to protect against identity theft by an attacker impersonating an individual by mimicking his or her biometric traits.

It is natural to draw a parallel between password-based authentication and biometric verification of identity. In a password-based system, a secret password is presented to confirm a claimed identity; in a biometric verification system, the trait is presented to confirm the claimed identity. It would appear that exposure of an individual's biometric data is comparable to disclosure of a secret password, with the added complication that while it is easy to replace a password, the same is not true for a biometric

---

[51]"Biometric technology is inherently individuating and interfaces easily to database technology, making privacy violations easier and more damaging. If we are to deploy such systems, privacy must be designed into them from the beginning, as it is hard to retrofit complex systems for privacy." Available at http://www.eff.org/Privacy/Surveillance/biometrics/.

[52]As noted in Chapter 1, access to sensitive systems should rely not just on the presentation of the correct biometric sample but rather on the security of the full process. Concerns about identity theft arise because not all biometric systems offer adequate security, and some could be vulnerable to attack by impersonators.

trait. Further, biometric data are exposed not only when data leak from unencrypted or poorly protected databases—they can, at least in principle, be derived from publicly observable human traits. The submission of a password and the presentation of a biometric trait are not, however, analogous. As discussed in Chapter 1, the security value of a biometric verification system stems from measures surrounding the presentation and capture of the biometric trait. These measures cope with public disclosure of an individual's biometric data by verifying that a presented trait is genuine and not an artifact employed by an attacker. However, when the sample capture is remote and unattended, as would be the case for most systems associated with computer access, there are few technical safeguards and minimal protection against the use of artifacts. In these circumstances, one would not expect a biometric recognition system to provide reliable protection against a premeditated attack.

SUMMARY

Although biometric systems can be beneficial, the potentially lifelong association of biometric traits with an individual, their potential use for remote detection, and their connection with identity records may raise social, cultural, and legal concerns. Such issues can affect a system's acceptance by users, its performance, or the decision on whether to use it in the first place. Biometric recognition also raises important legal issues of remediation, authority, and reliability, and, of course, privacy. Ultimately, social, cultural, and legal factors are critical and should be taken into account in the design, development, and deployment of biometric recognition systems.

5

Research Opportunities and the Future of Biometrics

The first four chapters of this report explain much about biometric systems and applications and describe many of the technical, engineering, scientific, and social challenges facing the field. This chapter covers some of the unsolved fundamental problems and research opportunities related to biometric systems, without, however, suggesting that existing systems are not useful or effective. In fact, many biometric systems have been successfully deployed. For example, hand geometry systems serve to control access to, among others, university dorms, nuclear power plants, and factories, where they record time and location.[1] Automated fingerprint identification systems (AFISs) integrate automatic and manual processes in criminal justice applications and civilian applications such as national identity systems.

An emerging technology such as biometrics typically confronts unrealistic performance expectations and is sometimes unfairly compared with approaches such as passwords that are not really alternatives. An effective biometric solution does not have to be—nor can it be—100 percent accurate or secure. For example, if there exists a 1 percent possibility of successful "buddy punching" (signing in for a friend or colleague), a hand geometry system can easily be seen as preventing 99 percent of such

---

[1]A.K. Jain, R. Bolle, and S. Pankanti, eds., *Biometrics: Personal Identification in Networked Society,* Norwell, Mass.: Kluwer Academic Publishers (1998), as cited in A.K. Jain, S. Pankanti, S. Prabhakar, L. Hong, A. Ross, and J. Wayman, *Biometrics: A Grand Challenge,* Proceedings of the 18th International Conference on Pattern Recognition, Cambridge, England (2004). Available at http://biometrics.cse.msu.edu/Publications/GeneralBiometrics/Jainetal_BiometricsGrandChallenge_ICPR04.pdf.

fraud. A particular application demands not perfection but satisfactory performance justifying the additional investments needed for the biometric system. In any given case, the system designer should understand the application well enough to achieve the target performance levels.

Nevertheless, solutions to the problem of recognizing individuals have historically been very elusive, and the effort needed to develop them has consistently been underestimated. Because humans seem to recognize familiar people easily and with great accuracy, such recognition has sometimes incorrectly been perceived as an easy task. Considering that a number of governments around the world have called for the nationwide use of biometrics in delivering crucial societal functions such as passports, there is an urgent need to act. Excepting for their application in national forensic AFISs, biometric recognition systems have never been tried at such large scales nor have they dealt with the wide use of nonforensic sensitive personal information. The current performance of some biometric systems—in particular with regard to the combination of error rate, robustness, and system security—may be inadequate for large-scale applications processing millions of users at a high throughput rate.

If there is a pressing public need for these applications, and if it is determined that biometric systems and technologies are the most appropriate way to implement them, then our understanding of the underlying science and technology must be robust enough to support the applications.[2] There is no substitute for realistic performance evaluations and sustained investment in research and development (R&D) to improve human recognition solutions and biometric systems.[3] The rest of this chapter outlines a research agenda focusing on (1) technical and engineering considerations, (2) social challenges, and (3) broader public policy considerations. The chapter concludes with a high-level overview of what constitutes a well-designed biometric system.

TECHNOLOGY AND ENGINEERING RESEARCH OPPORTUNITIES

In recent years several research agendas for biometric technologies and systems have set important challenges for the field.[4] The issues and

---

[2]The National Science Foundation Center for Identification Technology Research is one program taking an interdisciplinary approach to research related to biometrics. More information about the center is available at http://www.nsf.gov/eng/iip/iucrc/directory/citr.jsp.

[3]Standardization efforts, discussed elsewhere in this report, can help facilitate the cycle of build-test-share for transitioning the technology from concept to business solution.

[4]See, for example, A.K. Jain, S. Pankanti, S. Prabhakar, L. Hong, A. Ross, and J. Wayman, *Biometrics: A Grand Challenge*, Proceedings of the 18th International Conference on Pattern

research opportunities raised in this chapter are meant to complement, not replace or supersede these other articulations. Indeed, the U.S. government has created or funded several interdisciplinary, academia-based research programs that provide an institutional foundation for future work. The focus of this report has been on broad systems-level considerations, particularly for large-scale applications, and the technical challenges outlined in this Chapter reflect that focus. But as these other agendas demonstrate, there are numerous opportunities for deeper understanding of these systems at almost every level. This section lays out several technical and engineering areas the committee believes would benefit from sustained research and further investigation: human factors, understanding the underlying phenomena, modality-related technical challenges, opportunities to advance testing and evaluation, statistical engineering aspects, and issues of scale.

Human Factors and Affordance

Because biometric technologies and systems are deployed for human recognition applications, understanding the subject-technology interface is paramount. A key piece of the biometric recognition process is the input of the human characteristic to be measured. With the exception of recent work at NIST-IAD[5] and at Disney, very little effort has been expended on the "affordance"—the notion that what is perceived drives the action that occurs, or, put another way, that form can drive function—of biometric systems.[6] Biometric systems should implicitly (or explicitly) suggest to the user how they are to be interacted with.[7]

Recognition, Cambridge, England (2004); National Science and Technology Council, Subcommittee on Biometrics, "The National Biometrics Challenge" (2006), available at http://www.biometrics.gov/Documents/biochallengedoc.pdf; E. Rood and A.K. Jain, *Biometrics Research Agenda*, report of an NSF Workshop (2003); and Mario Savastano, Philip Statham, Christiane Schmidt, Ben Schouten, and Martin Walsh, Appendix 1: Research challenges, *BioVision: Roadmap for Biometrics in Europe to 2010*, Astrid Albrecht, Michael Behrens, Tony Mansfield, Will McMeechan, and Marek Rejman-Greene, eds. (2003), available at http://ftp.cwi.nl/CWIreports/PNA/PNA-E0303.pdf.

[5]M. Theofanos, B. Stanton, and C. Wolfson, Usability and biometrics: Assuring successful biometric systems, NIST Information Access Division (2008); M. Theofanos, B. Stanton, C. Sheppard, R. Micheals, J. Libert, and S. Orandi, Assessing face acquisition, NISTIR 7540, Information Access Division Information Technology Laboratory (2008); and M. Theofanos, B. Stanton, C. Sheppard, R. Micheals, Nien-Fan Zhang, J. Wydler, L. Nadel, and W. Rubin, Usability testing of height and angles of ten-print fingerprint capture, NISTIR 7504 (2008).

[6]J.J. Gibson, The theory of affordances, in *Perceiving, Acting, and Knowing*, Robert Shaw and John Bransford, eds. Hillsdale, N.J.: Erlbaum Associates (1977).

[7]This idea can be extended even to systems where the subject is unaware of the interaction but behaviors are being suggested to facilitate data collection.

International standards, such as the International Organization for Standardization (ISO)/IEC 19795 series,[8] generally contain an informative annex on best practices for data collection for the modality under consideration. For example, ISO/IEC 19794-5 contains an annex on best practices for face images specifying that full frontal face poses should be used and rotation of the head should be less than +/– 5 degrees from frontal in every direction. This requirement presents an affordance challenge that has not yet been adequately addressed by the technologies—namely, how can a system be designed to suggest to the user a pose that meets this requirement? (The committee was told about one system that presented an image that, when viewed from the proper angle, was clearly visible to the user.) Similar challenges exist with every modality/application combination and will require a modality- and application-specific set of solutions.

"Quality" has been used to indicate data collected in compliance with the assumptions of the matching algorithms, such that recognition performance of the algorithm can be maximized, which means that "affordance" and data "quality" are tightly linked. System operators and administrators face their own challenges when interfacing with the systems. How should the interfaces of attended systems be designed so that the operator knows how and when to collect proper images, how to recognize when poor-quality images have been collected, and how to guide the data subject in making better presentations? Very little research in this area has been conducted, and there is opportunity for significant progress.

Distinctiveness and Stability of Underlying Phenomena

There are many open questions about the distinctiveness of the underlying biometric traits in these systems and about human distinctiveness generally. One typical assumption in the design of most biometric systems has been that characteristics, if properly collected, are sufficiently distinctive to support the application in question. This assumption has not, however, been confirmed by scientific methods for specific biometric characteristics, either by prospectively collecting and analyzing biometric samples and feature patterns or by exploiting databases of samples or feature patterns assembled for other purposes. A broad and representative sampling of the population in which distinctiveness is being evaluated should be obtained and a minimum quality specification should be set to which biometric samples should conform. More generally, the development of a scientific foundation for reliably determining the distinctive-

---

[8]"Information technology—biometric performance testing and reporting," ISO/IEC 19795.

ness of various biometric traits under a variety of collection modes and environments is needed.

In other words, what is the effective limit on accuracy for a specific biometric trait in a realistic operating environment? This becomes a particularly important question at scale—that is, when the systems are expected to cope with large user populations and/or large reference databases. Even in DNA analysis, there has been controversy and uncertainty over how to estimate distinctiveness.[9] In biometric systems, "ground truth"—the collection of facts about biometric data subjects and recognition events to allow evaluation of system performance—is challenging, particularly for passive surveillance systems, where failures to acquire may be difficult to detect.[10]

There are also open questions about the stability of the underlying traits—how persistent (stable) will a given individual's biometric traits be over time? Some biometric traits, such as fingerprints, appear to be reasonably stable, but others, such as facial characteristics, can change significantly over even short periods of time. Depending on the capture and matching algorithms, changes in a trait over time may or may not have an effect on system performance and whether that person is appropriately recognized. Understanding more about the stability of common biometric traits will be important, especially if biometric systems are deployed for comparatively long (years or decades) periods of time.

All of this suggests several avenues of research that could strengthen the scientific underpinnings of the technology. There needs to be empirical analysis of base-level distinctiveness and the stability of common biometric modalities, both absolutely and under common conditions of capture, and research into what types of capture and what models and algorithms produce the most distinguishable and stable references for given modalities. Further, the scalability of various modalities under different capture and modeling conditions must be studied. The individuality of biometric identifiers, their long- and short-term physiological/pathological variations, and their relationship to the user population's genetic makeup all merit attention as well.

---

[9]P.J. Bickel, Discussion of "The evaluation of forensic DNA evidence," *Proceedings of the National Academy of Sciences* 94(11): 5497 (1997). Available at http://www.pnas.org/content/94/11/5497.full?ck=nck.

[10]Ted Dunstone and Neil Yager, *Biometric System and Data Analysis: Design, Evaluation, and Data Mining*, New York: Springer Science+Business Media (2008).

Modality-Related Research

Every biometric system relies on one or more biometric modalities. The choice of modality is a key driver of how the system is architected, how it is presented to the user, and how match vs. nonmatch decisions are made. Understanding particular modalities and how best to use the modalities is critical to overall system effectiveness. Research into several interrelated areas will bring continued improvement:

- *Sensors.* Reducing the cost of sensor hardware; improving the signal-to-noise ratio, the ease of use and affordability, and the repeatability of measures; and extending life expectancy.
- *Segmentation.* Improving the reliability of identifying a region of interest when the user presents his or her biometric characteristics to the system—for example, locating the face(s) in an image or separating speech signal from ambient noise.
- *Invariant representation.* Finding better ways to extract invariant representation (features) from the inherently varying biometric signal—that is, what kind of digital representation should be used for a face (or fingerprint or other feature) such that the trait can be recognized despite changes in pose, illumination, expression, aging, and so on.
- *Robust matching.* Improving the performance of the matching algorithm in the presence of imperfect segmentation, noisy features, and inherent signal variance.
- *Reference update.* Developing ways to update references so that they can account for variations and the aging of reference data in long-lived systems.
- *Indexing.* Developing binning and partitioning schemes to speed up searches in large databases.
- *Robustness in the face of adversaries.* Improving robustness to attacks, including the presentation of falsified biometric traits (perhaps, for example, through automated artifact detection).
- *Individuality.* Exploring the distinctiveness of a particular biometric trait and its relationship to the matching performance. Does information about distinctiveness serve to increase understanding of the effective limits on matching performance, for example?

In addition to the general challenges described above, there are also challenges specific to particular biometric modalities and traits. While the following discussion does not describe all the challenges for each modality, it does offer some potentially fruitful avenues of investigation for the most common ones.

An ongoing challenge for facial recognition is segmentation—distinguishing facial features from surrounding information. Another significant challenge for it is invariant representation—that is to say, finding a representation that is robust and persistent even when there are changes in pose, expression, illumination, and imaging distance, or when time has passed.

Specific challenges with respect to fingerprints include reducing the failure to enroll (FTE) and failure to acquire (FTA) rate, perhaps through the design of new sensors, artifact detection, image quality definition and enhancement, and high-resolution fingerprint matching. Fingerprint-based biometric systems could also be improved by increasing the speed of capture and minimizing contact, particularly for 10-print systems.

Iris recognition systems present R&D opportunities in the following areas: sensors; optimization of the illumination spectrum; reducing FTE and FTA rates; capturing and recognizing the iris at greater distances and with movement of the subject; and reducing the size of the hardware.

Improving speaker separation, normalizing channels, and using higher-level information (that is, beyond basic acoustic patterns) would all offer opportunities to improve voice recognition. In addition, robustness and persistence are needed in the face of language and behavioral changes and the limited number of speech samples.

Information Security Research

In many applications, biometric systems are one component of an overarching security policy and architecture. The information security community is extensive and has long experience with some of the challenges raised by biometric systems, which gives it a real opportunity for fruitful and constructive interaction with the biometrics community. Biometric systems pose two kinds of security challenges. The first is the use of biometrics to protect—provide security for—information systems. For what types of applications and in which domains is an approach incorporating biometric technologies most appropriate? This is a question for the broader information security community as well as the biometrics community and requires that we understand the goals and needs of an application to ascertain whether a biometrics-based approach is useful.

Assuming that a biometrics system is in place, the second security challenge is the security, integrity, and reliability of the system itself.[11] Information security research is needed that addresses the unique problems of biometric systems, such as preventing attacks based on the presentation of fake biometrics, the replay of previously captured biometric samples, and the concealment of biometric traits. Developing techniques

---

[11]See NRC, *Toward a Safer and More Secure Cyberspace*, Washington, D.C.: The National Academies Press (2007) for an in-depth discussion of security.

for protecting biometric reference information databases to avoid their use as a source of fake biometrics is another area for such research. Decision analysis and threat modeling are other critical areas requiring research advances that will allow employing biometric systems more fully across a range of applications.

Testing and Evaluation Research

Testing and evaluation are an important component in the design, development, and deployment of biometric systems. Several areas related to the testing and evaluation of biometric systems are likely to prove fruitful. This section describes a few of them. While there has been significant work on testing and evaluating a variety of approaches,[12] it is the committee's view that an even broader approach has merit. Moreover, while standardized evaluations of biometric systems are highly useful for development and comparison, their results may not reliably predict field performance. Methods used successfully for the study and improvement of systems in other fields (for example, controlled observation and experimentation on operational systems guided by scientific principles and statistical design and monitoring) should be used in developing, maintaining, assessing, and improving biometric systems. (See Chapter 3 for lessons that may be applicable from other domains.) The work over the last decade within the international standards community to reach agreement on fundamental concepts, such as how error rates are to be measured, has clarified the application of test methods under the usual laboratory conditions for biometric systems deployments.[13] Guidance for potential deployers of biometric systems on what is even a useful and appropriate initial set of questions to ask before getting into the details of modalities and so forth, as developed by a number of groups, has

---

[12]NIST's emerging National Voluntary Laboratory Accreditation Program (NVLAP) for biometrics represents progress in formalizing testing programs but does not yet provide specific testing methods required for different products and applications and does not yet address operational testing. The NVLAP Handbook 150-25 on Biometrics Testing is available at http://ts.nist.gov/Standards/Accreditation/upload/NIST-Handbook-150-25_public_draft_v1_09-18-2008.pdf.

[13]Note that while these standards are aimed at a broad swath of systems, they are not seen as appropriate for governments to use for large-scale AFIS systems, where they would need to be tested for conformance to standards, compliance with system requirements, alignment with capacity and accuracy requirements, and satisfaction of availability and other traditional system parameters.

proven particularly useful.[14] And of course, in addition to the technical questions that need to be addressed, there are issues regarding how to measure cost over the life cycle of the system and how to assess potential and actual return on investment (ROI). Unfortunately, ROI analysis methodologies and case studies have been lacking in comparison to other types of assessments. See Box 5.1 for a brief discussion and example of an ROI assessment.

Ultimately, determining the performance of an operational system requires an operational test, because adequately modeling all of the factors that impact human and technology performance in the laboratory is extremely difficult. Although the international standards community has made progress in developing a coherent set of best practices for technology and scenario testing, guidelines for operational testing are still under development and have been slowed by the community's general lack of experience with these evaluations and a lack of published methods and results.[15] Designing a system and corresponding tests that can cope with ongoing data collection is a significant challenge, making it difficult for a potential user of biometric systems, such as a federal agency, to determine how well a vendor's technology might operate in its applications and to assess progress in biometric system performance. Careful process and quality control analysis—as distinct from traditional, standardized testing of biometric systems that focuses on match performance for a test data set—at all stages of the system life cycle is essential. In addition, testing methods and results should be sufficiently open to allow disinterested parties to assess the results.

Test Data Considerations

One challenge meriting attention is test data for biometric systems. Designing large-scale systems requires large test data sets that are representative of the subject population, the collection environment, and system hardware expected in the target application. How does one determine which user population will be representative of the target application? The committee believes it is unlikely that being representative of the target application is the same as being representative of the population as a whole, because the population that should be considered will vary depending upon the ultimate application for which the system is

---

[14]For example, the recommendations of the U.K. Biometrics Working Group in "Use of Biometrics for Identification and Authentication: Advice on Product Selection, Issue 2.0" (2002). Available at http://www.cesg.gov.uk/policy_technologies/biometrics/media/biometricsadvice.pdf.

[15]ISO/IEC 19795-6, Biometric Performance Testing and Reporting—Part 6: Testing Methodologies for Operational Evaluation, is under development by ISO JTC1 SC37.

used. Legal and privacy concerns have limited the collection and sharing of both test and operational data (for example, various data sets collected by the U.S. government) with researchers,[16] raising the question of whether biometric data can be made nonidentifiable back to its origin.[17] If it cannot, could synthetic biometric data be created and used in lieu of real biometric data?[18] If the latter is possible, does the use of synthetic (imagined) data offer any scientific validity in assessing performance of a system using real data? When test results are available, who has access to them? These and related questions merit attention from not just the T&E community but the broader biometrics communities as well.

Usability Testing

Many factors related to usability can affect system effectiveness and throughput and may also affect how well the system performs its recognition tasks. Testing and evaluation mechanisms are therefore needed that provide insight into how well a system under consideration handles a variety of user interface expectations.

Despite the recent focus of NIST's information access division on usability testing, there is still major work to be done. One potential area of investigation is to incorporate into the design of the interface information on the expected motor control and cognitive capabilities of the user popu-

---

[16]Government operational biometric data—that is, personally identifiable information (PII)) for research and testing are governed by the privacy impact assessments (PIAs) and system-of-record notices (SORNs) associated with the specific systems, which are required by the Privacy Act of 1974. Whether the Privacy Act provides the latitude to use operational biometric and biometric-related data for large-scale research and testing purposes (during acquisition and operation) so long as data privacy and integrity are adequately protected is subject to interpretation. Various scenarios for how such data might be shared include these: (1) using the data (such as from IAFIS or US-VISIT) internal to the agency collecting the data, (2) using such data outside the agency collecting the data (such as providing multiagency data to NIST for analysis), or (3) providing such data to a university and/or industry team for analysis, etc. Assuming the Privacy Act permits such uses, then the PIAs and SORNs would have to specify such use of the data. The NIST multimodal biometric application resource kit (MBARK) data set is an example of test data collected but not disseminated for privacy reasons. The Department of Homeland Security (DHS) has released its former IDENT database to NIST for testing (NIST reports refer to it as DHS-2), but it has not been released to general researchers.

[17]Fingerprint images that are sufficiently similar to the original fingerprint can be reconstructed from data representations of the fingerprints. See J. Feng and A.K. Jain, FM model based fingerprint reconstruction from minutiae template, *International Conference on Biometrics* (2009), pp. 544-553. Available at http://biometrics.cse.msu.edu/Publications/Fingerprint/FengJain_FMModel_ICB09.pdf.

[18]SFINGE, a fingerprint synthesis technique, is described in Chapter 6 of D. Maltoni, D. Maio, A.K. Jain, and S. Prabhakar, *Handbook of Fingerprint Recognition*, 2nd edition, Springer Verlag (2009).

BOX 5.1
Return on Investment and Suitability Considerations

Determining the return on investment (ROI) for a biometric system is very much dependent on the application. It is based, among other things, on the risk the system is mitigating, the severity of the risk (projected loss in dollars should a security breach occur), and the anticipated benefits to the implementer of success. Making the business case for biometrics has proved difficult for many reasons, including the following: (1) the business value of security and deterrence—if they are the goal of the biometric system under consideration, is always difficult to quantify, regardless of technology; (2) fraud rates and costs of long-standing business systems (for example, PINs and passwords) are not well understood; and (3) total costs for biometrics systems have not been well documented or reported. Some media reports have been critical of biometric systems on the issue of return on investment, but not enough systematic study has been done on this issue to reach any firm, general conclusions.

A primary distinction between types of applications is between commercial applications aimed at, say, convenience or fraud reduction and applications whose goal is improving national or large organization security. Commercial applications of biometric systems are almost completely driven by financial considerations. Decisions about the implementation of commercial biometric systems are made based on expected cost savings, enhanced customer service or convenience, or regulatory compliance. When biometric systems are deployed as a component of a security apparatus, they face similar challenges to investment in cybersecurity—that is to say, the improvements are notoriously hard to quantify.

Commercial application ROI computations tend to be problematic, but those for national security applications tend to be even more difficult, with the biggest problems being determining the probability of an attack, the costs of a successful attack, and the life-cycle costs of the biometric system, including supervision and management costs. This is an area where case studies and research are badly needed and requires considering all the factors discussed in earlier chapters. It is especially important to take into account the trade-offs for life-cycle costs associated with

---

lations. Such information would allow the use of public health statistics to estimate the percentage of the general population (or subpopulation) that would be expected to have either cognitive or physical difficulties using the systems. By incorporating this understanding of the skills expected of users, designers and developers could tune the interfaces in ways that would increase their usability.

Usability is affected by other factors as well. For example, some unknown percentage of the population has a condition in which the fingers do not possess the usual friction ridges central to the functioning of fingerprint-based biometric system. In addition, some unknown (but believed to be nonzero) percentage of the population has either no irises or irises of unusual shape. When setting baseline error rates, it is important

equipment, maintenance, operational labor, speed of operation and impact on users (for example, on employees and travelers) both in absolute terms but also relative to the cost of doing nothing or of implementing another solution. It is important to know how the performance and trade-offs of the system can be characterized as well as which trade-offs are appropriate for which applications.

It may be that Disney is the first commercial biometric system deployer known to have experienced a complete life cycle, and their experience serves as a useful brief case study of ROI analysis. The company has gone through an end-of-life system analysis and made a business case decision to replace its existing finger geometry technology with a different, newer biometric modality and technology, multispectral fingerprint scanning. The company determined through internal testing and analysis that the newer technology could increase both the accuracy and throughput to a degree that provided an attractive ROI by cutting down on staff and increasing guest satisfaction.

In the Disney application, the objective was purely commercial, the risk probabilities and potential losses could be closely estimated, and the benefits could be measured directly. The objective was to implement a biometric to tie customers to their nontransferable gate passes in order to cut down on fraud (illegal pass resale or transference), which had grown to sizable proportions. Initially, the project applied only to season passes, which were issued simply as paper documents that could easily be transferred or resold, often by tour-group operators, who would transfer them (in violation of state law—but hard to enforce), costing Disney millions of dollars in lost revenues.

Once the season passes were tied to a biometric characteristic, which ensured that the user of the pass was the person who had originally purchased it,[1] the incidence of fraud fell dramatically and revenues grew commensurately. The ROI for the implementation of this biometric system was very easily calculated and turned out to be very high.

---

[1]The Disney system does allow transference of passes within families and other small affinity groups. This raised difficult design issues and has led some users of the system to wrongly conclude, after transferring passes between family members, that the system does not work.

to have estimates of the percentage of the population lacking the required trait, because this lack interacts with the design of sensors and algorithms. It may well be that each modality will have lower error rate bounds that cannot be improved upon by better sensors, algorithms, or collection procedures. More research is needed to understand this. Such questions are related to the distinctiveness and stability of the underlying biometric traits, discussed above.

Other usability considerations relate to the ease of participation. Is the system designed to take into account user needs (such as tables on which to set their items if necessary) and physical differences (such as height and weight)? What kinds of user assistance should be provided? What accommodations could be made for people who are unfamiliar with the system, or, conversely, for people who are very familiar with it (much as toll pass transponder users can use dedicated lanes on highways)?

ROI Analysis Methodologies and Case Studies

Determining the potential ROI and identifying which system characteristics contribute is an important means of evaluating any biometrics deployment. In addition to how well a system meets its requirements, there are issues about measuring cost over the life cycle of the system and assessing potential (and actual) ROI. There are relatively few ROI analysis methodologies and case studies. The research opportunity here is to develop methods for examining likely costs and cost savings that take into account the technical life cycle as well as ongoing maintenance and usage costs.

Evaluative Frameworks for Potential Deployers

In addition to system and technology tests, there is a significant opportunity to develop an evaluative model that would guide potential procurers and users of biometric systems. Guidance for potential users of biometric systems on an appropriate initial set of questions to ask before getting into the details of modalities and so forth has proven particularly useful.[19]

Testing When Data Changes

Designing a system and tests that can cope with ongoing data collection after it has been deployed is a significant challenge. The characteristics of the data may change from what was assumed during testing. This could be due to changes in the technology, changes in the user population, changes in how the system is used, or all of the above. Such potential changes to the data make it a challenge for potential users of biometrics, such as federal agencies, to readily determine how well a given vendor's technology might operate for the agency's applications over time. Developing testing and evaluation methodologies that can account for such potential changes or offer information on how a system's performance

---

[19]The NRC report *Protecting Individual Privacy in the Struggle Against Terrorists: A Framework for Assessment* (The National Academies Press, Washington, D.C., 2008) provides a framework for assessing program efficacy as well as impacts on privacy. The U.K. Biometrics Working Group developed another such document, which is available at http://www.cesg. gov.uk/policy_technologies/biometrics/media/biometricsadvice.pdf.

might change in the event of significant changes to the data characteristics is an opportunity for further investigation.

Operational Testing

Finally, operational testing is problematic in that most existing systems do not retain the data needed to determine error and throughput rates. Each system collects and stores different data in application-specific ways. Additionally, ground truth (all of the relevant facts about all participants) cannot be known in real applications with arbitrary user populations. Privacy rights of the data subjects may prevent using collected data for testing purposes. Lastly, because system operators may not wish others to know about operational performance for reasons of security, very few operational test results have ever been published. The ISO/IEC JTC1 SC37 standards committee has been working for a number of years on basic guidance for operational testing, but progress on this standard, to be known as ISO/IEC 19795, Part 6, has been slow, reflecting the inherent difficulties in making general statements about operational tests.

Systems-Level Statistical Engineering Research

In addition to the modality-related technical challenges outlined above, there are broad systems-level considerations to take into account. In particular, statistics and statistical engineering offer opportunities for progress and the development of design principles and model designs for operational testing of biometric systems and experimentation with modifiable system parameters. This section outlines some potential research questions in statistical engineering and biometric systems that merit attention.

Statistical approaches come into play with respect to the user populations, including cross-sectional and longitudinal studies of the variability of various biometric modalities over time, the association of biometrics with demographic and medical factors, and the effects of demographic factors and physical characteristics on failure to acquire and error rates.

Another systems-level consideration is error rates in biometric systems, including the following topics:

- The relative contribution to error of different phases and components of biometric recognition (on an algorithm-by-algorithm basis, because error rates and their causes are algorithm specific);
- The potential for algorithm-specific quality control measures to reduce error rates in varying populations of data subjects;
- The application of known statistical methods for analysis of cor-

related data to estimation[20] of false match and false nonmatch rates for recognition tasks; and

• The investigation of new statistical models for the estimation as above but from biometric databases where information on replication is incomplete—that is, replication is known only for a subset, or some information about variation in replicate measurements is available from an external source.

Other areas of potential investigation include studies of statistical efficiency and cost-benefits of different approaches to choosing, acquiring, and utilizing multimodal biometrics of various sorts. Issues to be considered would include the relative algorithmic-dependent within-class and between-class variation of sample choice such as multiple instances of 2 fingerprints versus single instances of 10 fingerprints versus single instances of 2 fingerprints combined with two-dimensional facial imaging. The aim is to develop the most informative fusion methods based on application- and algorithm-dependent multivariate distributions of biometric features.

Research on Scale

There are many ways in which scale may manifest itself in biometric systems. These include the number of sensors in the system, the number of comparisons being performed for a given unit time or a given input sample, the number of users (including administrators and operators), the geographic spread of the system, the size of the potential user population, or any combination of these factors. Addressing issues of scale in biometric systems offers numerous opportunities for research.

For instance, one question is this: How does the number of persons who have references in the enrolled database affect the speed of the system and its error rates? For some applications and associated algorithmic approaches, the size of the database might not matter if typical operation involves only a one-to-one comparison—that is, one set of submitted samples being compared to one set of enrollment records. For large-scale identification and screening systems, sequentially performing a very large number of one-to-one matches is not effective; there is a need for efficiently scaling the system to control throughput and false-match error rates as the size of the database increases.

Typical approaches to scaling include (1) using multiple hardware

---

[20]The committee notes the reluctance to use interval estimation because of lack of agreement on how to handle systematic errors except through personal probability. See ISO/IEC, Guide 98-3 (1995) *Guide to the Expression of Uncertainty in Measurement*.

units, (2) coarse pattern classification (for example, first classifying a fingerprint into classes such as arch, tented arch, whorl, left loop, and right loop), and (3) extensive use of exogenous data (such as gender, age, and geographical location) supplied by human operators. Although these approaches perform well in practice, they come at a price. Using hardware linearly proportional to the database size is expensive. Coarse pattern classification offers substantial scaling advantages even when single measures are available and even more advantage with multiple measures—for example, fingerprints from multiple fingers—but can add to the nonmatch error rates. Use of exogenous information creates a mechanism for intentionally avoiding identification—for example, dressing as the opposite sex or appearing older—if someone is trying to avoid being recognized by the system, not to mention potential privacy compromises.

Ideally, one would like to index biometric data patterns in some way similar to that used in conventional databases in order to benefit from lessons learned in other arenas. However, due to large intraclass variation in biometric data caused by variation in collection conditions and human anatomies and behaviors, it is not obvious how to ensure that samples from the same pattern fall into the same index bin. There have been very few published studies on reliably indexing biometric patterns.[21] Efficient indexing algorithms would need to be developed for each technology/modality combination. It is unlikely that any generic approach would be applicable to all biometric measures, although efforts to understand similarities and where lessons from one type of system can be applied to another are warranted.

False-match errors generally increase with the number of required comparisons in a large-scale identification or watch-list system. As most comparisons are false (for example, a submitted sample compared to the enrollment pattern of another person), increasing the size of the database increases the number of opportunities for a false match. However, in large-scale systems it is unlikely that a sample would be compared against every possible match in the database. Instead, just as with search algorithms generally, the set of items to compare against is winnowed according to certain criteria as quickly as possible to save time and memory. Because of the nonindependence of sequential comparisons using the same sample data, coupled with architectural and algorithm design choices that are aimed at finding any matches while sustaining throughput rate and limiting active memory, the relationship between the number

---

[21]J.L. Wayman, Multi-finger penetration rate and ROC variability for automatic fingerprint identification systems, in N. Ratha and R. Bolle, eds., *Automatic Fingerprint Recognition Systems*, New York: Springer Verlag (2003).

of false matches and database size is a poorly understood issue meriting further investigation.[22]

Although a watch-list database in a screening system is much smaller than that in a large-scale identification application, the number of continuous or active comparisons may be huge. Therefore, as in large-scale applications, the throughput and error-rate issues are also critical in screening applications. Computationally, scaling of large systems for near-real-time applications involving 1 million identities is becoming feasible, as is screening the traffic for 500 recognized identities. However, designing and building a real-time identification system involving 100 million identities is beyond our understanding. More research is needed here as well.

Social Science Research Opportunities

Biometric systems require an intimate association between people and the technologies that collect and record their biological and behavioral characteristics. This is true whether the application is overt or covert, negative claim or positive claim. It is therefore incumbent on those who conceive, design, and deploy biometric systems to consider the cultural and social contexts of these systems. Unfortunately, there are few rigorous studies of these contexts. Below is a framework for developing a portfolio of future research investigations that could help biometric systems better cope and perform within their cultural and social contexts.

Cultural and social issues arise at essentially two different levels—for the individual and for society. At the level of the individual, whether they are interacting actively or passively with a biometric system (for example, the person seeking entry to a facility), the issue is the performance of a biometric system. At the societal level, the issue is the social impact of the biometric system (for example, all are affected, either directly or indirectly, by the trade, tourism, and terrorism effects of a biometric passport).

At the level of the individual, social considerations are critical in the design, deployment and functioning of biometric systems. As we have noted, system performance may well be degraded if relevant social factors are not adequately taken into consideration. For example, religious beliefs that call for adherents to cover their faces in public make facial-recognition biometrics problematic. Thus if a biometric system is to work well for a broad range of people it must take into account behaviors resulting from

---

[22]See, for example, J.L. Wayman, Error rate equations for the general biometric system, *IEEE Robotics and Automation* 6(1): 35-48, and H. Jarosz, J.-C. Fondeur, and X. Dupré, Large-scale identification system design, J.L. Wayman, A.K. Jain, D. Maltoni, and D. Maio, eds., *Biometric Systems: Technology, Design, and Performance Evaluation*, New York: Springer (2005).

such things as religion or social convention. Every biometric system has a protocol for how it is to be interacted with. The protocol may be simple or complex, uniform in application, or tailored to the individual. Obviously, however, a good protocol for a biometrics system must recognize variations in biological features. A system based on fingerprints must have ways to gracefully accommodate a person who is missing a finger or who otherwise does not have usable fingerprints.

In addition to the design issue of affordance, previously discussed, research is needed to determine effective, appropriate, and graceful protocols, processes, and devices that gain the cooperation of participants, and the protocols and devices must be acceptable to the community. In biometric systems that are essentially surveillance systems, compliance should be thought of as more than acquiescence and should extend to gracefully (perhaps without notice) promoting the types of behavior (for example, face pose and angle) that result in useful biometric measures. Full compliance represents the ideal interaction of the participant with the biometric system from the viewpoint of the system designers. Acceptability to the community refers to the endorsement, or at least the lack of active disapproval, by significant governmental and public leadership groups. In any case, community acceptability is not guaranteed. Influential parts of the community may find biometric systems overly intrusive, unfair to certain groups, or inadequately protective of the individual's privacy.

The dimensions of individual compliance and community acceptability are discussed next. One part of the design process for particular systems or, more realistically, for a particular class of systems might be to develop data that predict how well the biometric system will perform in a target community and on factors that may make the system more acceptable to that community. Predictive aspects may just have a statistical relationship with subject compliance or community acceptability, while acceptability factors probably have a causal relationship. Developing data in these areas will provide the evidence needed to assess the relationships.

The extent and nature of participant compliance can be discovered and confirmed using either or both of two basic research strategies: field studies using ethnographic tools such as in cultural anthropology or attitude studies of using survey methods such as are common in sociology. Some of the things that might predict participant compliance include participant attitudes toward authority, their willingness to try new technology, their adherence to certain religious or cultural beliefs, and the geographic distribution of the population. Such work could, in theory, be part of the design work for biometrics systems but is generally not done, possibly because of the expense and effort involved. Research that sheds light on these issues would provide valuable information for those designing and building biometric systems.

Factors that motivate participant compliance can be discovered by experimental studies, essentially creating laboratory environments in which the factors can be controlled. This research paradigm is common in experimental psychology, but the extent to which such controlled studies might develop data that reflect factors encountered in operational applications is only speculative. Some candidate factors include self-interest, enforcement, inducement, social pressure, conviction, habit or practice, behavior of other actors, pleasantness of the experience, and attention to cultural norms. The more common approach is to survey data subjects who have just encountered an operational system to elicit their opinions,[23] but even this approach has rarely been applied.

Aspects that predict the extent and nature of community acceptability can be discovered and confirmed using either or both of two basic research strategies: field studies of similar deployments using ethnographic tools—as indicated for participant compliance above, or focus groups that are asked to discuss how they view various characteristics of a biometric system such as are common in marketing studies. The kinds of aspects that might be predictive of community acceptability include resemblance to existing well-tolerated systems, operated under the auspices of a respected institution, or a system that meets all legal requirements

One factor in motivating community acceptability is whether community concerns—for example, fairness, privacy, and confidentiality—are addressed. Using data from research in these and related areas, it should be possible to address a variety of relevant questions, such as: Where on the scale of purely voluntary to mandatory is a particular biometric system? In largely voluntary systems does cooperation vary by subgroup such as age, sex, or race? Does habituation lead to greater cooperation? How important is it that participants believe they or others will not be harmed? What factors influence such trust? What are effective and appropriate compliance mechanisms for biometric systems? Although it is not reasonable to expect designers of a specific system to conduct such research, these questions could be addressed as part of a more general research agenda.

---

[23]As was done, for example, by the Orkand Corporation in *Personal Identifier Project: Final Report*, California Department of Motor Vehicles report DMV88-89 (1990).

PUBLIC POLICY CONSIDERATIONS AND
RESEARCH OPPORTUNITIES

Numerous issues come into play beyond technical and engineering considerations in government use of biometric systems. These other issues include the following:

• To what extent can the need for a biometric system be satisfied by current technical capability? Balancing mandates with maturity of systems and technology is critical. Aggressive schedules can push technology development forward but not all challenges can be addressed on short notice.

• Is there sufficient flexibility and time to support the risk management needed to develop and deploy a biometric system? Governments must avoid increasing risk through overly constrained integration and testing timelines and budgets. The risks include the possibility not only that the system will fail or be compromised but also the possibility that the system will be rejected by its users or be so cumbersome or inefficient that it is withdrawn from use.

• Should participation in the system be mandated? Such a mandate might foster a climate of distrust or social unrest.

• What is the nature of the biometrics workforce? To the extent that biometric systems and related technologies are seen as important to meet public policy goals, is there sufficient incentive to grow and maintain the needed expertise? Training and maintaining consistent biometrics workforce has been difficult. Several organizations recently announced plans to create certification programs for professionals, but consensus must be reached on what skills are required of a professional in this area. The creation of a biometrics undergraduate program at West Virginia University is a step in the right direction. (The program has a ready customer: the FBI's Criminal Justice Information Center in Clarksburg, West Virginia.)

• The sourcing of the technology is crucial to the government's successful deployment of technological and information systems, including biometric systems. There is an inherent danger in relying on companies with manufacturing, research, or development activities centered overseas. For biometric systems, especially, the risk is the potential for U.S. biometric data to be collected by foreign governments, inviting scrutiny of U.S. information on border control systems and other critical infrastructure by persons not cleared by the U.S. government.

The social science considerations described above may have impacts on broader public policy considerations. Systematic empirical research and factual analysis would help provide an evidence case for public policy in this area. Some key research questions that have an impact on public policy include the following:

- What lessons can be learned from environmental impact statements and privacy impact assessments that might be relevant to deciding whether social impact assessments for biometric systems are useful?
- Do existing or proposed biometric systems represent a serious potential for identity theft?
- How have authoritarian regimes made use of human recognition methods to assert their control over individuals? In what ways might biometric systems enable these sorts of uses? How could such a risk be mitigated?
- To what extent are privacy requirements, interagency control issues, and policy constraints, or the perception thereof, inhibiting the research use and sharing of existing biometric data?
- What belief sets, if any, lead to an aversion to certain biometric technologies?

A reliable and effective biometric system may be perceived as providing irrefutable proof of identity of an individual, notwithstanding the many uncertainties already mentioned, raising concerns for users. Will the information regarding biometrics-based access to resources be used to track individuals in a way that infringes on privacy or anonymity? Will biometric data be used for an unintended purpose: For example, will fingerprints provided for access control be matched against fingerprints in a criminal database? Will data be used to cross-link independent records from the same person—for example, health insurance and grocery purchases? How would a user be reassured that a biometric system is being used for the intended purpose only?

Designing information systems—not only biometric systems—whose functionality is verifiable during deployment is very difficult. One solution might be a system that meticulously records recognition decisions and the people who accessed the logged decisions using a biometric-based access control system. Such a system could automatically warn users if a suspicious pattern is seen in the system administrator's access of users' logs. Another solution might be biometric cryptosystems—cryptographic keys based on biometric samples. Radical approaches such as total transparency attempt to solve the privacy issues in a novel way. But there are no obviously satisfactory solutions on the horizon for the privacy problem. Additional research on the relationship between biometric (bodily) information and privacy is needed.

The privacy protections required to facilitate data collection from and about biometric systems need to be clearly established. Because many of these systems are deployed to satisfy security needs, it is reasonable to expect that performance and vulnerability data need to be protected. For best results, the data sets for such research should be very large, contain very few errors in ground truth (metadata indexing), be appropriately randomized, and represent the populations of interest to target applications. To the extent consistent with privacy and security, the results of the studies should be published in the peer-reviewed scientific literature and the biometric samples used made widely available to other researchers.

REALIZING A WELL-DESIGNED BIOMETRIC SYSTEM

Research in the areas described above is warranted more than ever as biometric systems become widespread and are used in critical applications. This report concludes by taking a step back and presenting a vision of a well-designed biometric system that should persist even as progress is made on the challenges described earlier in this chapter.

A well-designed biometric system includes more than technology. It is a complex combination of technology, public policy, law, human processes, and social consensus. In the long term, there may be new modalities that allow recognizing human characteristics and behaviors quickly and effectively with little or no interaction on the individual's part. Human beings may turn out to possess distinctive traits that have yet to be fully explored or that cannot be suitably represented by present technology. Some of the potential sources of suitable signals currently being pursued include inductive signature and brain waves (EKG activity). Each of these potential signal sources could bring with it a new set of societal and policy issues requiring exploration.

Even with all of these uncertainties, and even with the many intriguing open questions that merit research, the committee believes that the following framework for a well-designed biometric system will apply for the foreseeable future. Progress in such research will lead to even more well-designed systems. This framework is offered as both an evaluative tool and as a development tool.

A well-designed biometric recognition system will have (at least) the following characteristics:

• The system will be designed to take into account that no biometric characteristic is entirely stable and distinctive. In other words, it will take into account that biometric similarity represents a likely, not a definitive, recognition and and that the corresponding is true for a failure to find similarity. In particular, presumptions and burdens of proof will be designed conservatively, with due attention to the system's inevitable imperfections.

• The policies of such a system will recognize that any claimed probabilities of correctness depend on external assumptions about dynamic presentation distributions, and that these assumptions, whether subjective or based on estimates from past internal or external data, are fallible. It will enable system operators and users to recognize that biometric information has a life cycle. Biometric information is collected or modified during, for example, enrollment, recognition, and so on. But policies should also recognize that changes in the biometric characteristics of the individual can lead to incorrect or failed recognition.

• The system will be designed so that system operators and users recognize that some inaccurate information may be created and stored in the databases linked to biometric references, and that over time information in these databases will become out of date. In particular, the reliability of information in the database is independent of the likelihood of correct recognition. The system will be designed to handle challenges to the accuracy of database information in a fair and effective way.

• Because the system's sensors and back-end processes are not perfectly accurate, it will need to handle failures to enroll, failures to acquire samples, and other error conditions gracefully and without violating dignity, privacy, or due process rights.

• Because some individuals will attempt to force the system into failure modes in order to avoid recognition, the system's failure modes must be just as robustly designed as the primary biometrics-based process.

• The system's security, privacy, and legal goals must be explicit and publicly stated, and they must be designed to protect against a specific and enumerated set of risks. The system will specifically address the possibility that malicious individuals may be involved in the design and/or operation of the system itself.

• It will recognize that biometric traits are inherently not secret and will implement processes to minimize both privacy risks and risks of misrecognition arising from this fact.

CONCLUDING REMARKS

This report lays out a broad systems view and outlines many of the subject areas with which biometrics research intersects. The committee also describes many open research problems, ranging from deep scientific questions about the nature of individuality to vexing technical and engineering challenges. It raises questions about appropriate system architecture and life-cycle design as well as questions about public policy regarding both private sector and government use of biometric systems. It notes that biometrics is an area that benefits from analyzing very large amounts of data. These and other aspects of biometrics suggest many fruitful areas and interesting problems for researchers from a range of disciplines.

Appendixes

A

Biosketches of Committee Members and Staff

COMMITTEE MEMBERS

JOSEPH N. PATO, *Chair,* is Distinguished Technologist at Hewlett-Packard's HP Laboratories. Previously he served as chief technology officer for Hewlett-Packard's Internet Security Solutions Division. He is currently a visiting fellow with the Massachusetts Institute of Technology Computer Science and Artificial Intelligence Lab Decentralized Information Group (DIG). Since 1986, Mr. Pato has been involved in security research and development, studying authentication, identification, and privacy issues. Currently Mr. Pato is developing a research program that will analyze security issues in the health care industry. Mr. Pato's current research focuses on the security needs of collaborative enterprises on the Internet, addressing both interenterprise models and the needs of lightweight instruments and peripherals directly attached to the Internet. Specifically, he is looking at critical infrastructure protection and the confluence of trust, e-services, and mobility. These interests have led him to look at the preservation of Internet communication in the event of cyberterrorism, trust frameworks for mobile environments, and applying privacy considerations in complex systems. His past work includes the design of delegation protocols for secure distributed computation, key exchange protocols, interdomain trust structures, the development of public- and secret-key-based infrastructures, and the more general development of distributed enterprise environments. Mr. Pato is also a founder of the IT Information Sharing and Analysis Center (IT-ISAC), where he has served as a board member. Mr. Pato has participated on several standards or advisory committees of the Institute of Electrical and Electronics Engineers (IEEE), American National Standards Institute (ANSI), National Institute for Standards and Technology (NIST), Department of Commerce, W3C, Financial Services Technology Consortium (FSTC), and Common Open Software Environment (COSE). He has represented Hewlett-Packard to the Open Software Foundation (OSF) Architecture Planning Council, the technical arm of the OSF Board of Directors. He has also served on the Technical Planning Committee evolving the Distributed Computing Environment (DCE) and chaired the Security and Remote Procedure Call (RPC)/Programming Model/Environment Services working groups. He has served as the vice-chair for the Distributed Management Environment (DME)-DCE-Security working group of the OSF Security Special Interest Group. In the past, Mr. Pato served as the co-chair for the OASIS Security Services Technical Committee, which developed Security Assertions Markup Language (SAML) from June 2001 until November 2002. SAML 1.0 was approved as an OASIS standard on November 1, 2002. Mr. Pato served as a key member of the NRC committee that wrote *Who Goes There? Authentication Through the Lens of Privacy* (2003). Mr. Pato's graduate work was in computer science at Brown University.

BOB BLAKLEY is vice president and research director for Burton Group Identity and Privacy Strategies. He covers identity, privacy, security, authentication, and risk management. Before joining Burton Group, Dr. Blakley was chief scientist for security and privacy at IBM and served on the National Academy of Sciences' Committee on Authentication Technologies and Their Privacy Implications. He has served as general chair of the 2003 IEEE Security and Privacy Conference and as general chair of the New Security Paradigms Workshop. Dr. Blakley is the former editor of the Object Management Group Common Object Request Broker Architecture (OMG CORBA) security specification and authored *CORBA Security: An Introduction to Safe Computing with Objects*, published by Addison-Wesley. He is also editor of Open Group's Authorization Application Programming Interface (API) specification effect and currently holds more than 10 patents on security-related technologies. Dr. Blakley received an A.B. in classics from Princeton University and a master's degree and Ph.D. in computer and communications sciences from the University of Michigan.

JEANETTE BLOMBERG is research staff member and program manager for practice-based service innovation at the IBM Almaden Research Center. Before assuming her current position, Dr. Blomberg was a founding member of the pioneering Work Practice and Technology group at the

Xerox Palo Alto Research Center (PARC), Director of Experience Modeling Research at Sapient Corporation, and industry-affiliated professor at the Blekinge Institute of Technology in Sweden. Since joining IBM Research she has led projects on interactions among IT service providers and their clients, collaboration practices among globally distributed sales teams, the place of stories in corporate imaginaries, and new approaches to work-based learning. Her research explores issues in social aspects of technology production and use, ethnographically informed organizational interventions, participatory design, case-based prototyping, and service innovation. Dr. Blomberg is an active member of the participatory design community, having served as program co-chair twice, and she sits on a number of advisory boards, including the Foresight panel of the IT University of Copenhagen, the Program in Design Anthropology at Wayne State University, and the Ethnographic Praxis in Industry Conference (EPIC). Dr. Blomberg received her Ph.D. in anthropology from the University of California, Davis; before embarking on her career in high tech, she was a lecturer in cultural anthropology and sociolinguistics at UC Davis.

JOSEPH P. CAMPBELL received B.S., M.S., and Ph.D. degrees in electrical engineering from Rensselaer Polytechnic Institute in 1979, the Johns Hopkins University in 1986, and Oklahoma State University in 1992, respectively. Dr. Campbell is currently a senior member of the technical staff at the Massachusetts Institute of Technology Lincoln Laboratory in the Information Systems Technology Group, where he conducts speech-processing research and specializes in advanced speaker recognition methods. His current interests are high-level features for speaker recognition and forensic-style applications of it, creating corpora to support speech-processing research and evaluation, robust speech coding, biometrics, and cognitive radio. Before joining Lincoln Laboratory, he served 22 years at the National Security Agency (NSA). From 1979 to 1990, Dr. Campbell was a member of NSA's Narrowband Secure Voice Technology research group. He and his teammates developed the first DSP-chip software modem and LPC-10e, which enhanced the Federal Standard 1015 voice coder and improved U.S. and NATO secure voice systems. He was the principal investigator for the CELP voice coder and led the U.S. government's speech coding team in developing for it. CELP became Federal Standard 1016 and is the foundation of digital cellular and voice-over-the-Internet telephony systems. From 1991 to 1998, Dr. Campbell was a senior scientist in NSA's Biometric Technology research group, where he led voice verification research. From 1994 to 1998, he chaired the Biometric Consortium, the U.S. government's focal point for research, development, test, evaluation, and application of biometric-based personal identification and verification technology. From 1998 to 2001, he led the Acoustics Section of NSA's Speech Research branch, conducting and coordinating research on and evaluation of speaker recognition, language identification, gender identification, and speech activity detection methods. From 1991 to 1999, Dr. Campbell was an associate editor of *IEEE Transactions on Speech and Audio Processing*. He was an IEEE Signal Processing Society Distinguished Lecturer in 2001. From 1991 to 2001, Dr. Campbell taught speech processing at the Johns Hopkins University. Dr. Campbell is currently a member of the IEEE Signal Processing Society's board of governors; an editor of *Digital Signal Processing* journal; a chair of the International Speech Communication Association's Speaker and Language Characterization Special Interest Group (ISCA SpLC SIG); a member of ISCA, Sigma Xi, and the Acoustical Society of America; and a fellow of the IEEE.

GEORGE T. DUNCAN is a professor of statistics, emeritus, in the H. John Heinz III College and the Department of Statistics at Carnegie Mellon University. He was on the faculty of the University of California, Davis (1970-1974) and was a Peace Corps volunteer in the Philippines (1965-1967), teaching at Mindanao State University. His current research work centers on statistical confidentiality. He has published more than 70 papers in such journals as *Statistical Science, Management Science, Journal of the American Statistical Association, Econometrica,* and *Psychometrika*. He has received National Science Foundation (NSF) research funding and has lectured in Brazil, Cuba, England, Italy, Turkey, Ireland, Mexico, and Japan, among other places. He chaired the Panel on Confidentiality and Data Access of the National Academy of Sciences (1989-1993), producing the report *Private Lives and Public Policies: Confidentiality and Accessibility of Government Statistics*. He chaired the American Statistical Association's Committee on Privacy and Confidentiality. He is a fellow of the American Statistical Association, an elected member of the International Statistical Institute, and a fellow of the American Association for the Advancement of Science. In 1996 he was elected Pittsburgh Statistician of the Year by the American Statistical Association. He has been editor of the Theory and Methods Section of the *Journal of the American Statistical Association*. He has been a visiting faculty member at Los Alamos National Laboratory and was the Lord Simon Visiting Professor at the University of Manchester. He received a B.S. (1963) and an M.S. (1964) from the University of Chicago and a Ph.D. (1970) from the University of Minnesota, all in statistics.

GEORGE R. FISHER runs both an investment advisory firm, George Fisher Advisors LLC, and a real-estate-development finance company, George Fisher Finance, LLC, in Boston, Massachusetts. He was formerly on the board of directors of Prudential Securities in New York City, where he led the merger of Prudential Securities into Wachovia. Prior to joining Prudential, Mr. Fisher was chief information officer at Fidelity Investments, where he managed technology oversight, consolidating mutual fund and brokerage platforms. Mr. Fisher also spent 16 years at Morgan Stanley, first as a principal for Technical Services Worldwide, transforming manual, low-volume systems into Wall Street leaders; later, as a managing director, he became the deputy general manager for Morgan Stanley Asia, restructuring Chinese operations and managing the explosive growth of Asia's regional markets. Mr. Fisher earned a B.A. in economics and computer science from the University of Rochester. He has also earned certifications from the National Association of Securities Dealers (Series 3, 7, 63, 24, and 27), the National Association of Corporate Directors (Director of Professionalism), the American Chamber of Commerce in Japan, and the Association of International Education of Japan, earning Level 4 language proficiency. Mr. Fisher is a certified financial planner (CFP).

STEVEN P. GOLDBERG was the James and Catherine Denny Professor of Law at the Georgetown University Law Center. An expert in law and science, Mr. Goldberg was the author of *Culture Clash: Law and Science in America* (1996), winner of the Alpha Sigma Nu Book Award, and coauthor of the widely used text *Law, Science, and Medicine*. He served as a law clerk to D.C. Circuit Court Chief Judge David L. Bazelon and U.S. Supreme Court Justice William J. Brennan, Jr. He also served as an attorney in the General Counsel's Office of the U.S. Nuclear Regulatory Commission. Mr. Goldberg was a member of the District of Columbia and Maryland bars and the Section on Science and Technology of the American Bar Association. He received an A.B. from Harvard College and a J.D. from Yale Law School. He died on August 26, 2010.

PETER T. HIGGINS, founder of Higgins & Associates, International, has 41 years' experience in the information technology field and has been involved with biometrics since the late 1980s. He was a member of the UK Home Office, Identification and Passport Service's Biometrics Assurance Group (UK BAG) from 2006 through June 2009, when UK BAG's mission was completed. He was an instructor of biometrics at the University of California, Los Angeles, Extension School for many years. He chaired the International Association for Identification's AFIS Committee for 5 years and is a well-known consultant and lecturer in the field of large-scale biometric procurement and testing. In 2002 he joined John Woodward and Nick Orlans in authoring the McGraw-Hill/Osborne book *Biometric Identification in the Information Age*. In 2004 he joined Peter Komarinski, Kathleen Higgins, and Lisa Fox in writing the Elsevier Academic Press book *Automated Fingerprint Identification Systems*. Previously he served as deputy assistant director of engineering with the FBI and was the program manager for the FBI's Integrated Automated Fingerprint Identification System (IAFIS). Before that, he served in technical, operational, and executive positions with the Central Intelligence Agency. Mr. Higgins received a B.A. in mathematics from Marist College and an M.S. in theoretical math and computer science from Stevens Institute of Technology in Hoboken, New Jersey.

PETER B. IMREY, a biostatistician and epidemiologist, is a professor of medicine at the Cleveland Clinic, where he holds appointments in the Cleveland Clinic Lerner College of Medicine of Case Western Reserve University, the Lerner Research Institute's Department of Quantitative Health Sciences, and the Neurological Institute's Mellen Center for Multiple Sclerosis Treatment and Research. He previously taught at the University of North Carolina (biostatistics, 1972-1975) and the University of Illinois (medical information science; statistics; and community health, 1975-2002), and is an adjunct professor, Department of Statistics, University of Illinois at Urbana-Champaign. Dr. Imrey has made research and expository contributions to statistical analysis of categorical data and diverse health science areas, including meningococcal disease, diet and cancer, and dental data analysis. He has served on editorial boards of three statistical journals and the *Encyclopedia of Biostatistics* (2nd edition), on numerous federal special study sections and emphasis panels, and on the National Academies' committee that produced the report *The Polygraph and Lie Detection* (2003). He has held multiple biostatistical leadership posts in the International Biometric Society (an international society promoting the development and application of statistical and mathematical theory and methods in the biosciences), the American Statistical Association, and the American Public Health Association (APHA). Dr. Imrey has been honored by APHA's Statistics Section and is a fellow of the American College of Epidemiology and a member of Sigma Xi and Delta Omega honorary societies. He received an A.B. in mathematics and statistics from Columbia University and a Ph.D. in biostatistics from the University of North Carolina at Chapel Hill.

ANIL K. JAIN is University Distinguished Professor in the Department of Computer Science and Engineering at Michigan State University. His research interests include pattern recognition, exploratory pattern analysis, Markov random fields, texture analysis, object recognition, and biometric authentication. He received the best paper awards in 1987, 1991, and 2005 and was cited for outstanding contributions in 1976, 1979, 1992, 1997, and 1998 from the Pattern Recognition Society. He also received the 1996 *IEEE Transactions on Neural Networks* Outstanding Paper Award. He is a fellow of the IEEE, the ACM, AAAS, SPIE, and the International Association of Pattern Recognition (IAPR). He was editor in chief of the *IEEE Transactions on Pattern Analysis and Machine Intelligence* (1991-1994). He has received a Fulbright Research Award, a Guggenheim fellowship, the Alexander von Humboldt Research Award, the IEEE Computer Society Wallace McDowell award, the King-Sun Fu Prize from IAPR, and the IEEE ICDM Outstanding Research Contribution award. He holds six patents in the area of fingerprint matching and has written a number of books on biometrics, including *Handbook of Biometrics*, *Handbook of Multibiometrics*, *Handbook of Face Recognition*, and *Handbook of Fingerprint Recognition*. He is a member of the Defense Science Board and served on the National Academies Committee on Improvised Explosive Devices. Dr. Jain was the co-organizer of the NSF workshop on the biometrics research agenda, held in May 2003, and has organized several conferences on biometrics. He received a Ph.D. in electrical engineering from Ohio State University in 1973.

GORDON LEVIN is senior engineer with the Advanced Systems group of Design and Engineering at Walt Disney World in Orlando, Florida, where the world's largest commercial biometric application has been operating since 1997. As a licensed electrical engineer, he is the engineer of record for all physical security system design performed on the 42-square-mile property. Mr. Levin has been a member of the Biometric Consortium Working Group (BCWG) since 1999 and the sole commercial end user to be a participating representative acting under NIST and the NSA to incubate biometric standards for submission to ANSI and the International Organization for Standardization (ISO). In 2002 he was the keynote speaker at the plenary meeting of ISO/IEC Joint Technical Committee 1 Subcommittee 37 on Biometrics. He also participated in the Aviation Security–Biometrics Working Group, which was assembled in the wake of 9/11 to report on passenger protection and identity verification. This report was instrumental in the strategic planning for the soon-to-be-formed Transportation Security Administration (TSA) and its plans for adopting biometric technology. Prior to joining Walt Disney World in 1997, Mr. Levin had been a private consultant engineer working in the DOD and commercial sectors in specialized security and electronic system design and construction.

LAWRENCE D. NADEL is a fellow in the Center for National Security and Intelligence, Identity Discovery and Management Division at Noblis. With more than 15 years' experience in biometrics, he has focused on the requirements for implementing effective and interoperable biometric systems and issues associated with it, objective methods for testing and evaluating the performance of these systems, and means for assessing biometric sample quality. Dr. Nadel currently leads Noblis's Biometrics Support Services Program for the Department of Homeland Security (DHS)/US-VISIT. He contributed to US-VISIT's conversion from 2-prints to 10-prints, helped strategize innovative solutions to biometric verification of non-U.S. citizens departing the U.S., and is supporting US-VISIT's implementation of multibiometrics and standards-based data sharing. He has provided technical leadership to other national identification and security-related projects for the FBI, TSA, DOD, and the National Institute of Justice. Dr. Nadel was co-principal investigator for a 2-year Noblis-funded research project investigating approaches to multi-biometric fusion. He has chaired the Noblis Biometric Identification Cluster Group since early 2000 and is a participant in the INCITS-M1 biometrics standards development group and the NIST NVLAP Biometrics Working Group. Dr. Nadel earned a B.S. in electrical engineering from the Polytechnic Institute of NYU and M.Sc. and Ph.D. degrees in electrical and biomedical engineering from the Ohio State University.

JAMES L. WAYMAN is a research administrator in the Office of Graduate Studies and Research at San Jose State University. He served as director of the U.S. National Biometric Test Center in the Clinton administration (1997-2000). He holds four patents in speech processing, is a principal U.K. expert on the ISO/IEC standards committee biometrics, and the editor of the ISO/IEC standard on voice data format. Dr. Wayman is a senior member of the IEEE and a fellow of the IET. He is coeditor of J. Wayman, A. Jain, D. Maltoni and D. Maio, *Biometric Systems* (Springer, 2005) and was a member of the NRC's committee that wrote *Who Goes There? Authentication Through the Lens of Privacy*. He has been a paid biometrics advisor to 8 national governments. Dr. Wayman received his Ph.D. degree in engineering from the University of California, Santa Barbara, in 1980.

STAFF

LYNETTE I. MILLETT is a senior program officer and study director at the Computer Science and Telecommunications Board, National Research Council of the National Academies. She currently directs several CSTB projects, including a comprehensive exploration of sustaining growth in computing performance and an examination of how best to develop complex, software-intensive systems in the DOD environment. She served as study director for the CSTB reports *Social Security Administration Electronic Service Provision: A Strategic Assessment* and *Software for Dependable Systems: Sufficient Evidence?* Her portfolio includes significant portions of CSTB's recent work on software, identity systems, and privacy. She directed the project that produced *Who Goes There? Authentication Through the Lens of Privacy*, a discussion of authentication technologies and their privacy implications; and *IDs—Not That Easy: Questions About Nationwide Identity Systems*, a post-9/11 analysis of the challenges presented by large-scale identity systems. She has an M.Sc. in computer science from Cornell University, where her work was supported by graduate fellowships from the NSF and the Intel Corporation; and a B.A. with honors in mathematics and computer science from Colby College, where she was elected to Phi Beta Kappa.

B

Watch-List Operational Performance and List Size: A First-Cut Analysis

Let p be the probability that someone presenting to a watch-list system has been previously enrolled, and $F(\cdot)$ be a prior distribution on this probability. $F(\cdot)$ may be discrete and even a point prior with all mass at one possible value π of p, a continuous distribution on the interval $[0,1]$ such as a Beta distribution, or any other probability distribution function on a probability space on $[0,1]$. Two types of results may be distinguished here: matching an enrolled presenter to the correct prior enrollment sample or, less restrictively, recognizing that the presenter has previously enrolled, although perhaps by matching to the wrong enrollee. The latter is pertinent to watch-list performance because such a result would serve the intended function of denying privileges, even if for the wrong reason. We distinguish here between these two possibilities by referring to the first as identification and to the second as watch-list recognition.

Addressing the identification problem first, one is trying to match a person specifically with his or her enrollment record and is in error if the correct match is missed. The confidence we should place in a claimed match—that is, its "predictive value"—is the probability that a claimed match is correct:

$$PPV(p) = P(\textit{true match with enrollment sample} \mid \textit{claimed match with enrollment sample}) =$$

$$\frac{P(\textit{true match with presenter's enrollment sample})}{P(\textit{claimed match with anyone's enrollment sample})} =$$

$$\frac{p \times P(\textit{true match} \mid \textit{enrolled})}{p \times P(\textit{any match} \mid \textit{enrolled}) + (1 - p) \times P(\textit{any match} \mid \textit{unenrolled})}$$

Consider the effect on this predictive value of enrolling one additional person in a watch list of length n, assuming the pattern of presentations to the list is fixed at proportion p of previous enrollees. In addition to comparisons with the slightly shorter previous list, the presenter is now compared to the new enrollee. This cannot increase and may decrease P(*true match | enrolled*), because each comparison offers an additional opportunity for an enrolled presenter to be erroneously matched with the wrong enrollee by matching more closely with someone else's stored data than with his or her own. Similarly, both denominator terms cannot decrease and may increase, because the new comparison offers any presenter an oppportunity of falsely matching with an extra enrollee.

Hence the ratio, $PPV(p)$, cannot increase and may decrease with watch-list length. Using the subscript to indicate watch-list length, $PPV_{n+1}(p) \leq PPV_n(p)$ for any specific p. Thus, the posterior means for the two list sizes over the distribution $F(p)$ must hold the same relationship:

$$E(PPV_{n+1}(p)) = \int_0^1 PPV_{n+1}(p)dF(p) \leq \int_0^1 PPV_n(p)dF(p) = E(PPV_n(p)).$$

These expectations are the marginal probabilities that a claimed match is correct for the different list sizes, so increasing list length by one enrollee cannot increase and may be expected to decrease the confidence warranted by a watch-list identification. Iterating this point shows that lengthening the list by any amount must have the same implications. However, this argument depends on decoupling the presentation distribution $F(p)$ from enrollee characteristics. In a finite population setting, where increasing enrollment increases p, a much more complicated argument might be required, with the outcome dependent on the specifics of functional relationships. A general argument that would work in such a setting is not obvious.

Our confidence in a nonmatch is $NPV(p) =$

P(unenrolled and claimed nonmatch | claimed nonmatch) =

$$\frac{P(unenrolled\ and\ claimed\ nonmatch)}{P(claimed\ nonmatch)} =$$

$$\frac{(1-p) \times P(claimed\ nonmatch\ |\ unenrolled)}{(1-p) \times P(claimed\ nonmatch\ |\ unenrolled)\ +}{p \times P(claimed\ nonmatch\ |\ enrolled)}.$$

As noted above, increasing watch-list size by one new enrollment without changing p offers an additional opportunity for each unenrolled presenter to falsely match. Thus, *P(claimed nonmatch | unenrolled)* cannot increase and may decrease. The new enrollee can affect results only for those enrolled presenters failing to match their enrollment samples and gives such presenters an additional chance to match the watch list, although incorrectly, thus decreasing $P(claimed\ nonmatch \mid enrolled)$. Assuming that list size does not affect the presentation distribution $F(p)$, the net impact depends on the ratio of the two probabilities. In the simplest conceivable model, when comparisons between pairs of individuals are independent and true and false-match probabilities are uniformly $\mu1$ and $\mu0$, these are respectively $(1 - \mu0)n$ and $(1 - \mu1)(1 - \mu0)^{n-1}$ when n subjects are enrolled, and both are multiplied by $(1 - \mu0)$ with each new enrollment, leaving their ratio and $NPV(p)$ unchanged. But if $\mu0$ depends on enrollment status, as might occur when attempts are made to compromise the identification process, then $NPV(p)$ can decrease or increase when $\mu0$ is higher for comparisons of unenrolled to enrolled presenters, or of one enrolled to other enrolled presenters, respectively. The expectation would change accordingly, in either direction.

Considering the watch-list recognition problem from the same perspective, one is now satisfied with a claim that the presenter matches someone on the list, without concern for whether the match is to the presenter's own enrollment sample. The definition and above discussion of NPV remain unaltered because a false match of a presenting enrollee, which is the event adjudicated differently by identification and watch-list recognition, does not contribute to probabilities conditioned on the absence of a match. Moreover,

$$PPV(p) =$$

$P(true\ match\ with\ any\ enrollment\ sample \mid claimed\ match\ with\ list) =$

$$\frac{P(claimed\ match\ with\ list\ and\ true\ match\ with\ any\ enrollment\ sample)}{P(claimed\ match\ with\ list)} =$$

$$\frac{p \times P(claimed\ match\ with\ anyone \mid enrolled)}{\begin{array}{c} p \times P(claimed\ match\ with\ anyone \mid enrolled)\ + \\ (1 - p) \times P(false\ match \mid unenrolled) \end{array}}$$

With a new enrollment to the list, an enrolled presenter who fails to match the correct enrollment sample has an added chance of matching the new enrollee and being correctly flagged as previously enrolled. This increases the numerator probability rather than decreasing it, as was the case for individual identification: numerator and denominator thus both increase. In the simple case described above, PPV can be shown to decline with list size, as was the case for identification. However, other scenarios and results are conceivable; if match probabilities differ for enrolled and unenrolled presenters, the prior distribution $F(\cdot)$ depends on list size, and match comparisons may be dependent.

For an example of how linkage of $F(p)$ to list size can change these results, consider a closed set identification system scaled up by enrolling many more users, each of whom interacts with the system daily to obtain workplace access, perhaps in a rapidly expanding corporation. Unless the number of attempted intrusions increases greatly, $F(p)$ is shifted to the right and p stochastically increases. In the resulting change, the increasing dominance of the *PPV* fraction by its numerator term outweighs the increasing chance of false recognition for any single impostor challenge, because impostor challenges occur with declining relative frequency. Confidence in a match would thereby increase rather than decrease.

C

Statement of Task

The Computer Science and Telecommunications Board (CSTB) will convene an expert committee to provide a comprehensive assessment of biometrics that examines current capabilities, future possibilities, and the role of government in their development. This project will build on CSTB's recent project on authentication technologies and privacy. It would explore the technical and policy challenges associated with the development, evaluation, and use of biometric technologies and the systems that incorporate them. It would examine associated research challenges and identify a multi- and interdisciplinary research agenda to begin to meet them. Multiple stakeholders and points of view on multiple technologies, applications, and implementation issues would be examined as part of the study. International perspectives and considerations would be explored as well. Throughout the study, inputs would be gathered through testimony and written material on the challenges, capabilities, and requirements of biometric systems as well as related policy and social questions.

D

Testing and Evaluation Examples

Large-scale biometric systems traditionally undergo a series of tests beyond technology and scenario testing. These large-scale system tests are typically at the system level, not just the biometric subsystem level, and occur multiple times in the life of a system in such forms as factory acceptance tests before shipment, site or system acceptance tests before initiating operations, and in-use tests to ensure that performance remains at acceptable levels and/or to reset thresholds or other technical parameters.

The following examples outline aspects of testing from three real-world, large-scale systems: the FBI's Integrated Automated Fingerprint Identification System (IAFIS), Disney's entrance control system, and the U.S. Army's Biometrics Automated Toolkit (BAT). Note that there are numerous other systems that offer lessons as well—these include the DOD's Common Access Card, US-VISIT, California's Statewide Fingerprint Imaging System (CA SFIS), and so on. The inclusion or exclusion of systems in this discussion is not meant to convey a judgment of any sort.

THE FBI'S IAFIS SYSTEM

IAFIS underwent numerous tests before and after deployment in 1998 and 1999. After deployment the FBI tracked performance for 5 years to determine the threshold for automatic hit decisions. After measuring and analyzing the performance data, the FBI was able to say with confidence that all candidates for whom the 10-fingerprint to 10-fingerprint search score was above a certain level were always declared as matches (or hits) by the examiners. As a result of this analysis the FBI was able to let the IAFIS system automatically declare as matches about 25 percent of the hits that previously required human intervention. This test used routine operational data in an operational environment and was not orchestrated to include any controls or prescreening on the input data. The transactions were run through the system normally, and match decisions were made by human examiners working with candidates presented by the IAFIS automated matchers.

DISNEY'S ENTRANCE CONTROL SYSTEM

Walt Disney World (WDW) has publicly reported[1] on internal testing using several different biometric technologies over the years. (See Box 5.1 for more on Disney's use of biometrics.) WDW tested various hand geometry and finger scanning technologies at several theme park locations to evaluate alternative technologies to the then-existing finger geometry used in its turnstile application. WDW also tested technologies for other applications to increase guest service and improve operating efficiency. Testing there is done in four stages: laboratory testing, technology testing, scenario testing, and operational evaluation. Since WDW has had existing biometric technology in place since 1996 and a substantial amount of experience with the biometric industry, its mind-set is that a threshold has been set for performance in both error rates and throughput and prospective vendors must exceed this level of performance to be considered for future enhancement projects.

In WDW lab testing, prospective biometric devices or technologies are examined for the underlying strengths of their technology/modality, usability, and accuracy. This testing is performed under optimal, controlled conditions for all of the relevant parameters that can affect performance. Parameters like technology construction and architecture, component mean time between failures, and theoretical throughput are extrapolated based on the results of laboratory tests. The goal of laboratory testing is to quickly determine whether a device or technology is worth investigating further. If a technology does not meet a performance level above the existing technology under optimal conditions, there is no point in investigating further.

If a prospective biometric device or technology is determined to be promising in the WDW lab environment, then the next stage of testing,

---

[1]Available at http://www.biometrics.org/bc2007/presentations/Tues_Sep_11/Session_ III/ 11_Levin_IBIA.pdf.

called "technology testing," is conducted to examine the limitations of the technology where some of the parameters will be controlled and others will be allowed to vary into "extreme limits" to see how the technology reacts and where it fails. For example, increasing the amount of ambient lighting for a facial recognition system or increasing the amount of background noise for a voice system stresses the capabilities of those systems. If the technology is still determined to be promising, scenario testing is performed by testing the technology in the live, operational environment with real-world users.[2] Typically, all data are captured and subsequently analyzed to determine if the system performed as expected, better, or worse. Analysis is performed to determine if some parameter was unexpectedly affecting performance. Often video of the user interactions will be recorded to assist in the data analysis and is particularly useful if the results of the testing show unexplained anomalies or unexpected results. For example, video of the interactions may detect users swapping fingers between enrollment and subsequent use in a fingerprint system. During this entire testing process, a potential system enhancement cost/benefit analysis is updated with the results of each round of the testing. If the performance gain is determined to be worthwhile, a business decision may then be made to migrate to the new technology. Disney has followed these scenario tests with operational tests on deployed systems to estimate actual error and throughput rates.

U.S. ARMY'S BIOMETRIC AUTOMATED TOOLKIT

BAT was developed in the late 1990s by the Battle Command Battle Lab of the Army Intelligence Center at Fort Huachuca, Arizona, as an advanced concepts technology demonstration (ACTD) to enable U.S. military forces to keep a "digital dossier" on a person of interest.[3] Other features such as biometric collection and identification badge creation were also included. BAT uses a rugged laptop and biometric collection devices (facial images, fingerprints, iris images, and, in some cases, voice samples) to enroll persons encountered by the military in combat operations. Hundreds of devices were rushed into production to meet demand during Operation Iraqi Freedom and Operation Enduring Freedom in Afghanistan.

This military use of biometrics ensures that a person, once registered, can later be recognized even if his or her identity documents or facial characteristics change. This permits postmission analysis to identify per-

---

[2]J. Ashbourn, "User Psychology and Biometric Systems Performance" (2000). Available at http://www.adept-associates.com/User%20Psychology.pdf.

[3]Available at http://www.eis.army.mil/programs/dodbiometrics.htm.

sons of future interest and in-mission analysis to detain persons of interest from biometrically supported watch lists. These systems are considered by many to be a technical success today, and the data are shared, when appropriate, with the FBI, DHS, and the intelligence community. When first deployed they did not go through factory or system acceptance tests due to the rapid prototyping and the demand for devices. After operational use, it was determined that the fingerprints collected were not usable by the FBI because several factors had not been considered in the original tactical system design, which did not include sending output to the FBI's strategic system, IAFIS. BAT was then formally tested operationally and the required changes identified and made. The operational retests before and after deployment showed that the current generation BAT systems generally met all of the image quality and record format protocols specified by the FBI. These BAT devices, however, use proprietary reference representations to share information on watch lists, which makes them less interoperable with standards-based systems than with one another.

E

The Biometrics Standards Landscape

Since September 11, 2001, there has been increased interest in using biometrics for national security purposes, some of which have been codified in legislation, including the Enhanced Border Security and Visa Entry Reform Act of 2002[1] and the PATRIOT Act of 2001.[2] As a result, biometric standards activities, previously largely limited to the forensics community, have been accelerated through national and international standards bodies. To speed up development of standards, NIST helped to establish a national standards body and requested the formation of an international standards body, both of which aim to increase the development and deployment of national and international biometrics standards for a variety of applications.

The following sections outline the main biometrics standards bodies, discuss some specific standards, and describe some of the challenges facing the processes. As with standards in other technologies, biometric standards face tension between being flexible enough to enable innovation while sufficiently prescriptive and detailed to allow interoperability and useful comparison of technologies and their capabilities.

---

[1]PL 107-173.
[2]PL 107-56.

STANDARDS BODIES

To facilitate standards work at the international level, ISO/IEU JTC 1/SC 37 was established in June 2002 at the request of the United States, which is represented within SC37 by the International Committee for Information Technology Standards, M1 technical committee on biometrics (INCITS M1). This body coordinates the development of biometric standards based on consensus development with the industrial, academic, and government communities. Within SC 37, there are six areas of focus: (1) vocabulary and concept harmonization; (2) biometrics data transfer; (3) data format standards for interoperability; (4) standard specific application profiles; (5) performance testing and reporting; and (6) cross-jurisdictional (legal and social) aspects for nongovernmental applications of biometrics.[3] INCITS M1 participates in five of these six working groups. Development of security standards for biometrics is specifically outside the remit of SC37 but is included in the work of SC27. Other international standards bodies include the International Telecommunication Union (ITU-T) and the International Civil Aviation Organization (ICAO). Specific work is also being carried out by specialized international groups including OASIS and the Open Group.

Nationally, the American National Standards Institute (ANSI) coordinates voluntary standardization and conformity assessments in the United States, approves the creation of all national and international standards, and may also implement any changes to the standards as well.

MAJOR STANDARDS

Standards aim to establish generic sets of rules for different products and to facilitate interoperability, data exchange, consistency of use, and other desirable features. One outcome of standards development is to achieve stability and consistency of biometric technologies and products that benefit consumers and investors. In the past decade, over 20 major international standards have been developed and approved, including the following:

- BioAPI, to enable hardware interoperability while retaining previous data,
- Fingerprint minutiae 19794-2,
- Fingerprint image 19794-4,
- Face image 19794-5,

---

[3]The work of WG6 was limited to nongovernmental applications at the request of INCITS M1.

- Iris image 19794-6,
- Hand geometry image 19794-10, and
- Testing and reporting fundamentals 19795-1.

ANSI has approved multiple national standards for the exchange of biometric data, two biometric application profiles, two biometric interface standards, and the Common Biometric Exchange Formats Framework. These standards are generally closely related to corresponding international standards, either serving as input to the creation of the corresponding international standard or simply repeating an established international standard. Application tensions can arise when the ANSI standard and the corresponding international standard have significant differences. The National Science and Technology Council Subcommittee on Biometrics and Identity Management released the report "Registry of USG-Recommended Biometric Standards Version 1.0" on June 5, 2008, clarifying which parts of which standards should be used for U.S. government applications. Two other standards are described below as examples of the challenges and complexities that can arise in the development of biometric standards.

Standard for the Interchange of Biometric Data

For the past 25 years the FBI and NIST have successfully developed and updated a standard for the exchange of fingerprint information. This standard is officially titled *Information Technology: American National Standard for Information Systems—Data Format for the Interchange of Fingerprint, Facial, & Other Biometric Information—Part 1 NIST Special Publication 500-271 ANSI/NIST-ITL 1-2007*. That standard has recently been updated to XML format to reflect the emerging needs of the defense and intelligence communities and defines about 20 record types for use in exchanging biometric information (faces, fingers, palms, latent prints, and irises) and related biographic and event data (e.g., date and time of enrollment). It has been one of the most successful and widely used biometrics standards. Criminal justice, border control, national identity, and social benefits programs all over the world have adopted the ANSI/NIST standard for the exchange of fingerprint images for their electronic fingerprint-based transactions.

This standard permits communities of interest, known as domains, to implement those portions of the standard that are relevant to their needs. For instance, the FBI's implementation, known as the Electronic Biometric Transmissions Specification, or EBTS, permits local, state, and federal agencies and departments to electronically exchange biometric and biographic information across various criminal-justice-oriented networks, independent of the source (vendor) of the equipment used. When the FBI first issued the EFTS[4] in 1994, it selected Type 4 high-resolution gray-scale images and stated that they would not accept Types 3, 5, or 6 fingerprint images. As a result, those record types are not used in any large-scale fingerprint automation projects.

After producing the EFTS the FBI next focused on image quality standards. Fingerprint image quality is the dominant factor in the AFIS ability to match fingerprints. The FBI added an Appendix F to the EFTS to specify image quality specifications (IQS) for capture devices and printers. This IQS standard is also used worldwide in procurement of live scan devices, fingerprint card scanners, and printers. Each domain (e.g., Interpol) has its own implementation document that specifies which records and which demographic data fields it will accept. These implementation documents normally show their relationship to the FBI's EFTS to include the IQS.

NIST developed an automated tool to rate the quality of fingerprint images. In October 2004 NIST released an updated version of this suite of tools for handling digital fingerprint images. NIST Fingerprint Image Software 2 was developed by NIST's Image Group for the FBI and DHS and is available free to U.S. law enforcement agencies as well as to manufacturers and researchers of biometric systems. New to this release is a tool that evaluates the quality of a fingerprint scan at the time it is made. Problems such as dry skin, the size of the fingers, and the quality and condition of the equipment used can affect the quality of a print and its ability to be matched with other prints. The tool rates each scan on a scale from 1 for a high-quality print to 5 for an unusable one.

NIST also worked with the FBI to develop fingerprint data compression standards acceptable to the latent print examination community. This compression standard, known as Wavelet Scalar Quantization (WSQ), is widely used in both forensic and civil AFIS systems, although newer systems seem to be migrating to use of JPEG-2000 for the compression of fingerprints. It is important to note that the 10-fingerprint images are compressed for transmission and storage while the latent print images are never compressed.

The FBI works with industry to permit vendors to self-assess products in order to place the products on the FBI certified products list. The self-certification reports are evaluated by FBI personnel supported by MITRE Corporation experts. Products on the list are typically certified as meeting Appendix F IQS with specific software drivers and operating system releases.

This is a case of a particular project or system, IAFIS, requiring a stan-

---

[4]Electronic fingerprint transmission specification (EFTS) is the predecessor of the current EBTS. The EBTS expands upon the EFTS to include additional biometric modalities.

dard and driving the development and implementation. Most of the other standards activity in the biometrics arena is driven not by a project but by a more general sense of a need for interoperability and a level playing field for technology providers.

Fingerprint Minutiae Exchange Standard

One area where there is a big push to develop and implement an extension to the ANSI/NIST standard for fingerprint exchange is the exchange of certain types of minutiae (or features) records rather than images. While the standard as written permits the exchange of minutiae in lieu of images, the minutiae defined in the standard are not as useful for processing across different vendor environments, from an algorithmic perspective, as permitting a vendor to receive an image and extract their proprietary minutiae set.[5] The ANSI/NIST standard currently supports at least eight vendor minutiae sets, per Table 15 of the standard.

The reasons for the push in this direction are twofold. First, when agencies exchange fingerprints for searching rather than retention, fingerprint minutiae can be transmitted and searched much more rapidly than fingerprint images. Second, when verifying the identity of a person presenting a personal identity verification (PIV) card, it would be time-consuming to extract the fingerprint image from the PIV card's chip and extract the features each time the card is used. Storing a common, interoperable set of minutiae on the card was selected to reduce transaction time considerably. The standard selected for storing templates on PIV cards is ANSI/INCITS 378.

Benchmarking is a form of testing often used in large-scale AFIS and Automated Biometric Identification Systems (ABIS)[6] source selection. In the 1980s there was an ANSI/IAI standard for Benchmarking AFIS Systems,[7] but when the time came to update the standard a decision was made to not update it; as a result, in conformance with ANSI process rules, the standard was allowed to fade away.

---

[5]A new standard for fingerprints that includes extended features is available at http://fingerprint.nist.gov/standard/cdeffs/Docs/CDEFFS_DraftStd_v03_Final.pdf. A recent paper shows there is a strong performance gain in using extended features for latent fingerprint matching. See Anil K. Jain and Jianjiang Feng, Latent fingerprint matching, *IEEE Transactions on Pattern Analysis and Machine Intelligence,* February 25 (2010). IEEE Computer Society Digital Library, IEEE Computer Society, available at http://doi.ieeecomputersociety.org/10.1109/TPAMI.2010.59.

[6]Automated Biometric Identification Systems are modeled on the function of AFIS systems but are not tied to finger imaging modalities and are often multimodal.

[7]American National Standard for forensic identification—automated fingerprint identification systems—benchmark tests of relative performance [ANSI/IAI 1-1988].

One trend in large-scale AFIS benchmarking is to perform three sets of tests that are described below. This approach is becoming an ad hoc standard for large-scale AFIS benchmarking:

- Operational demonstrations,
- Lights-out performance, and
- Best practices performance.

Operational demonstrations are intended to evaluate user interfaces, compression rates, scanner flexibility, end-to-end workflows, report generation, and administrative tasks. Lights-out testing measures the performance of the underlying biometric matchers for fingerprints, palm prints, and latent impressions with no human intervention other than feeding scanners and "lassoing" latent impressions within an image. Best practices performance testing measures the performance of the underlying biometric matchers for fingerprints, palm prints, and latent impressions, with fingerprint personnel permitted to perform quality control steps such as sequence correction and editing of low-quality images.

Another trend in benchmarking large-scale matcher systems that will be servicing larger systems is to bring the algorithms in house and run them under very controlled conditions against millions of records.

CHALLENGES IN THE BIOMETRIC STANDARDS ARENA

Despite the growing interest in and increasing approval of adopting biometric standards, a variety of challenges remain.

So-called patent ambush is one such challenge. It involves embedding a company's proprietary information in a standard and revealing the information only after the standard has been approved by a standards body, with the intention to exclude some companies from using the standard or to extract higher royalties from other companies that use the standard. Although proprietary information may become part of the standard, companies are required to formally disclose such information. However, the standards process should also uncover instances of patented technology as a proposed standard proceeds through review and approval phases. Instances of patent ambush have occurred in other technology industries[8] and are the subject of litigation in the area of biometrics.

As the standards process is designed to enhance the competitiveness of biometrics markets, many biometrics companies want to develop

---

[8]See, for instance, a discussion of patent ambush in telecommunications standards, available at http://www.lawdit.co.uk/reading_room/room/view_article.asp?name=../articles/EC%20Closes.htm.

their own standard rather than pay royalties to use another company's approved standard. This is not unique to biometrics by any means but often results in international standards bodies granting standards to the companies that propose them. A related challenge is that standards inevitably involve compromises and thus end up as a lowest common denominator among the various companies offering competing commercial biometric products. Evaluation and testing might then require more than mere standards compliance. NIST, for example, has conducted performance tests at a level that surpasses the standards that have been established by the international standards body. Two tests that have included additional criteria by NIST include the facial recognition challenge 2006 test and the MINEX 2006 test, which aim to enable interoperability of fingerprints at the minutiae level.

Interoperability presents its own problem in the standards arena. What is an appropriate or useful level of interoperability? How can we arrive at a shared definition? These issues have been addressed by the international standards community,[9] but work remains to be done. A related problem of interoperability is the tendency to decrease overall performance as the standard seeks the lowest common factors among the interoperating technologies. Multimodal biometrics fusion (MBF) can add more complexity to the standards process. MBF is the combining of more than one biometric modality, such as combining a fingerprint with an iris scan. (See Chapter 2 for more details.) Establishing standards for multimodal biometrics presents additional challenges, given the difficulties of establishing interoperability among unimodal biometrics. Many of these issues have been discussed in ISO/IEC documents.[10]

---

[9]Information technology—Biometric performance testing and reporting—Part 4: Interoperability performance testing, ISO/IEC 19795-4:2008.

[10]Information technology—Biometrics—Multimodal and other multibiometric fusion, ISO/IEC Technical Report 24722:2007.